Praise for *Unsettling: Surviving Extinction Together*

"In her love letter to the environment, Elizabeth Weinberg threads through geography, plant, and animal existence to dislocate and relocate some of our colonizing mythologies in order to reconnect humans to the planet. Moving through a queer, feminist lens, readers can begin to radically rethink the stories we tell ourselves about who we are. A triumph of ecological imagination."

—Lidia Yuknavitch, national bestselling author of
The Book of Joan and *The Chronology of Water*

"From the gigantic glory of whale poop to the humbling majesty of Denali, Elizabeth Weinberg's *Unsettling* is a wonderfully scientific and passionately personal exploration of the intertwining human disasters of climate change and lack of social justice. Required reading for anyone thinking we are the dominant species."

—Jon Scieszka, First US National Ambassador
of Children's Literature, and author of the climate-
activist graphic novel series *AstroNuts*

"Deeply researched and beautifully written, *Unsettling* deftly weaves memoir, history, and science to chronicle all we have lost: species, rivers, forests, Indigenous cultures and peoples, and our place in a living landscape. But in braiding Weinberg's own coming-out story with her profound connections to the world around her, *Unsettling* also presents a path away from an antagonistic relationship to the natural world and toward a fluid, equitable, and hopeful future."

—Maya Sonenberg, author of *Bad Mothers, Bad Daughters*
(winner of the Richard Sullivan Prize for Fiction)

"Facing the severe realities of climate change can be overwhelming. Elizabeth Weinberg's dazzling book crafts a welcome entry ramp, inviting readers on a deeply researched, wonder-filled, big-hearted exploration. She weaves science, history, social justice, and personal growth into elegant nature writing with a conscience. Weinberg is a compelling tour guide, deepening our connection to the planet and one another."

—Hannah Malvin, founder and director of Pride Outside

"'It's too late' are hopeless words that make us either fade back into the dissonance between humanity and the natural world or burn out in climate grief. *Unsettling* reminds us that beauty has not just existed but *persists*. Weinberg asks us what might happen if we meet each other's gaze, grieve what is lost, and resist the certain doom of 'It's too late,' to be the heroes we each need in what still is and what will be."

—Jenny Bruso, founder of Unlikely Hikers

"At once a celebration of our world, a Jeremiad, and a memoir, *Unsettling: Surviving Extinction Together* makes the stakes of climate change devastatingly clear. Whether she's illuminating the biodiversity of the site of a whalefall in the deep sea or the scale of the Cordilleran and Laurentide ice sheets or the traumatized resilience of urban coyotes, Elizabeth Weinberg is a passionate advocate for the ways in which we're going to have to reconceive our very relation to the planet in order to become our own heroes and turn aside oncoming catastrophe."

—Jim Shepard, author of *Like You'd Understand, Anyway* and *Phase Six*

UN
SETTLING

UN SETTLING

SURVIVING EXTINCTION TOGETHER

ELIZABETH WEINBERG

BROADLEAF BOOKS
MINNEAPOLIS

UNSETTLING
Surviving Extinction Together

Copyright © 2022 Elizabeth Weinberg. Printed by Broadleaf Books, an imprint of 1517 Media. All rights reserved. Except for brief quotations in critical articles or reviews, no part of this book may be reproduced in any manner without prior written permission from the publisher. Email copyright@1517.media or write to Permissions, Broadleaf Books, PO Box 1209, Minneapolis, MN 55440-1209.

Cover image: Landscape: Nastasic/iStock; coyote: SlothAstronaut/iStock
Cover design: Michele Lenger

Print ISBN: 978-1-5064-8205-7
eBook ISBN: 978-1-5064-8206-4

For Leslie

CONTENTS

INTRODUCTION

Off the coast of Oregon, a gray whale swims along the seafloor, bumping and scraping her head along the bottom of the ocean, loosening the soft sediment into a dense, waterborne cloud. She gulps in a mouthful of the muddy water, briefly savoring its earthy, salty taste, then forces it out through the hairy baleen hanging from her upper jaw. She scrapes the crustaceans left behind off with her tongue and slurps them down. She surfaces for a breath, then dives to feed again.

Off the coast of Oregon, that same whale logs at the surface, half asleep and breathing through the twin holes on the top of her head. She sighs, and steam rises into the air. Then, from the other end of her body, she lets out a mighty plume of digested invertebrates. All that protein has to go someplace, after all.

This whale friend of ours is massive, nearly fifty feet long. She's eaten more than a ton of food today. Her poop is prodigious.

Whales like her are so big, and so numerous, that their excrement drives entire ecosystems. Many species eat deep underwater, then defecate at the surface. Scientists describe their poop as *flocculent*, meaning wooly, like the tufts and curls adorning sheep. The loose, clumping nature of whale poop allows it to float, a mass of nutrients bobbing along the water like so many little sheep.

But it doesn't stay there for long. Bacteria and algae consume the nutrients and grow, then are eaten by small invertebrates and fish. Those, in turn, are eaten by larger fish and seabirds, which are then eaten by sharks, sea lions, and whales, until whale poop is fueling a whole web within the ocean. Scientists call it the whale pump.

It's just shit, sure. But as so often seems to happen in the natural world, the pieces that seem least significant are the most important.

And the whale, so often thought of as the tip-top of the food chain, also drives the bottom of it. Because why live a life that's just one thing?

———

Most adult baleen whales only eat for half the year. They spend the summers at high latitudes gorging themselves on krill and fish, building up their fat reserves, then journey thousands of miles to the tropics or subtropics to mate or give birth over the winter. The lower latitudes have fewer predators, making them a safe place to rear a vulnerable newborn, but they are also, for the most part, devoid of food. During these sun-warmed winters, whales will fast for months on end.

For several years, I worked as a science writer for a system of marine protected areas in the United States, including one that protects the breeding and calving grounds of thousands of humpback whales in Hawai'i. Most of my work was remote, and I rarely got into the field. But when I did get to go to Hawai'i, my spouse Leslie and I tacked on a vacation, including a day of whale watching and snorkeling. A catamaran took us off the coast of Maui early in the morning, and we had been motoring for about half an hour when the captain cut the engine. In the stillness, a puff of fog rose into the air, and then, there they were—a mother humpback whale and her calf, resting together.

It was the first time I'd seen whales in person, and I wanted to dance and to cry. They were so *big*, even the baby. And despite all we'd put whales through—centuries of whaling, noise pollution, climate change—they were here, and they were together.

That was in 2017, and though I didn't realize it at the time, seeing that mother and calf was even more miraculous than it felt. Beginning in 2015, fewer whales began showing up in Hawai'i each year, and those who were spotted were mostly males. In 2017, the Hawai'i Marine Mammal Consortium only sighted six calves in its annual survey of the Big Island's Kohala Coast.

Researchers are still trying to figure out why the whale numbers dropped, but it seems to be linked to food availability: thanks to a combination of climate change and El Niño, conditions in the North Pacific were particularly hostile to krill and other humpback whale food sources in those years. With less food in their Alaska feeding grounds, whales probably couldn't build up their fat reserves, and the females either skipped the breeding season and stayed up north to eat—or they died. Either way: fewer whales in Hawai'i.

If the whales couldn't build up their reserves, they couldn't reproduce, or possibly even survive. And that meant a deficit for everyone—for the small animals that depended on the whale poop up in Alaska, and for the males that came south in search of a mate. I'd seen a small miracle while a population was possibly in crisis, thanks to our human insistence on warming the world.

———

I wrote most of this book while the Trump administration was in power. Each day, the world seemed to be getting worse, and no one was being held accountable. Each day, I woke with a feeling of dread in the pit of my stomach, until finally, that dread gave way to numbness. But I, like so many people, still clung to a certain sort of certainty. I needed the world to get better, and so I believed it would. The administration would end, and we would have intelligent, reasonable people running the country again, and inevitably, the world would improve.

And to some degree, that came true: he's out of office. But I am writing this on a late summer day in Oregon during a hundred-degree heatwave when the light is hazy orange because millions of acres of the West are on fire. The Delta variant of COVID-19 is surging because people won't get vaccinated and won't wear masks. Earlier this week, the United Nations released a report that says that climate change is already impacting every inhabited region of the world, and even if we stop emissions right away, we're looking at catastrophe. I need things to get better, and I hope they will, but I also carry quite a bit of doubt. How do you radically change a world where extractive capitalism is the rule? How do you change the actions of eight billion people, each with their own agenda, whether it's getting rich or just surviving?

I want to learn to live as part of the world, not to feel like I'm on the sidelines, waiting for it to steady itself out of this endless wobble.

Against all these impending disasters, so much of what's in the world seems unimportant until you look closer. A whale's shit isn't just shit: it drives the ocean like a beating heart. I want to learn to take that closer look—before it's too late.

August 2021

PART ONE

"The climate crisis is also a crisis of culture, and thus of the imagination."

—Amitav Ghosh, *The Great Derangement*

1

GRAVEYARD

Picture an eighteen-wheeler with two trailers trucking down the highway, and you still haven't conjured the length of a blue whale. Pile together thirty African elephants, and you're still shy of its massive weight. The blue whale is the largest animal to exist on this planet, *ever*—even the heaviest dinosaurs were only about half its weight. A blue whale's body contains more than twenty billion miles of arteries, veins, and capillaries.

Whales—both blue whales and their smaller-yet-still-enormous cousins, like gray whales, humpbacks, and bowheads—are so large that once they reach adulthood, they don't usually have to worry about other animals eating them. They die of old age, of illness, of starvation, or because we've hit them with a container ship or entangled them in fishing gear and drowned them. We see evidence of their death when they perish near the shore, washing up along the coast, but if a whale dies out in the open ocean, it usually stays there. For a short time, it will bob along the surface; then it sinks. Resting on the bottom of the ocean, it transforms into a city.

Partially decomposed, the whale emits a ghastly array of putrid scents, beckoning to predators around it. Sharks and deep-sea fish journey miles, ready for a rare meal. Over the next year or two—or seven to ten if it's a particularly large species like a blue whale—these scavengers will devour the whale's fleshy nutrients and swim off, satiated, in search of their next foraging spot.

Once large predators have ripped apart the carcass, smaller creatures move in to take advantage of what is left. Crabs, lobsters, and other crustaceans wander over; polychaete worms and mollusks burrow in to eat their fill. Over time, this army of scavengers will strip the carcass down to the bone.

All these predators, large and small, leave the detritus of their meals behind on the seafloor. There, it decomposes further, creating a rich slurry of nutrition where before there was only mud. Bacterial mats and other microorganisms unfurl, metabolizing whatever they can find.

This takes years, and throughout it all, this graveyard is astonishingly biodiverse. If you were to walk four steps away from a whale fall in any direction and categorize all the creatures you found, you'd have more small organisms than in any other recorded habitat below a thousand meters. Coming upon a whale fall in the deep sea is like wandering through a paved parking lot, then plunging yourself into a patch of tropical rainforest.

Finally, once all the soft tissues are gone, it's time for the bones. Large whale bones are exceedingly fatty—a humpback whale skeleton may contain the weight of two small cars in fats—and that's a bonanza to any organism that can manage to get it. Specialized worms have evolved in tandem with whale falls for just this purpose. Tiny, feathered *Osedax* worms expertly burrow into the bone, dissolving the bone matrix with acid and extracting the fats. Bacteria play a role, too, converting sulfides within the whalebone into energy. The bones may last these organisms for up to a century, and even after that, the bones are essentially rocks, the tombstone of a whale waiting for a passing anemone or deep-sea coral larva to anchor to it like lichen.

More than four hundred species, from worms to sharks, are known to colonize whale falls, and many species were first discovered in this unlikely skeletal habitat. The death of a whale may be a tragedy, especially when humans are involved. But to the right organisms, it's also a rare oasis.

———

One of my first encounters with skeletons and graveyards was watching *The Lion King*. I saw the movie in theaters. I was five years old, a worrier and a wimp, and though the movie scared me, I was proud of myself for sitting through it, refusing to cry frightened tears even when I wanted to. I hugged my knees as Simba and Nala wandered into the elephant graveyard despite being warned away from this dangerous, disturbing place. I, too, was curious about what existed in that uncharted wilderness, but the scene where they run through heaps of bones chased by hyenas was as terrifying as it was enthralling.

The elephant graveyard is a desolate place. Mist and fog sweep across purple-gray cliffs while hulking bones pile in mountains like brambled hedges. Hyenas, portrayed as dirty, stupid scavengers, lurk in the gigantic skull of a once-powerful elephant. Nothing grows. It's a forbidden, liminal zone, the only part of the Pride Lands not ruled by the lions, and I dreamed of roaming through it myself, discovering its secrets.

After seeing the movie, I played Lion King with my best friend constantly. I always insisted on being Simba while she was Nala; I didn't want to be a girl and certainly not a secondary character. We would run across the perfectly groomed lawn of my parents' suburban backyard, two lion cubs scampering across the savannah and escaping the rigid conformity of adult supervision. Then we would enter the gloomy copse of trees that separated my parents' yard from our neighbors', hiding behind tree trunks and turning over rocks to see what weird creatures we could find. There in the elephant graveyard, we would face and miraculously escape certain doom.

———

Early in *The Lion King*, Mufasa takes Simba on a walk through the Pride Lands and tells him about the circle of life, the law of the lions' domain. While it may seem unfair that the lions both rule and eat the antelope, Mufasa explains that when lions die, they turn into grass, which the antelope eat. Everyone contributes; everyone gets their due.

Mufasa's explanation parallels a basic tenet of ecology: the importance of balance. Whatever you take out, you have to put back in some form for everything to function properly. If you don't, the environment will change until it finds a new equilibrium. The circle of life is one of the first times a Disney movie explicitly grapples with the idea of environmental balance and stewardship. It was an idea I'd carry forward throughout my childhood as I got to know different landscapes and ecosystems. Everything has its cycle; you just have to look for it.

When Mufasa's brother Scar takes over, everything goes to shit, demonstrating exactly what happens when you jam up the circle of life. He promotes the hyenas, formerly low-ranking scavengers, to the lions' equals, and the balance of this Disneyfied savannah is thrown off. There's no food; there's no water. The lionesses are forced to travel further and further in search of prey.

When Simba returns to the savannah, the Pride Lands are indistinguishable from the elephant graveyard. The movie shifts from lush greens and golds to moody purple; bones are everywhere; the river is dry.

We're creating our own elephant graveyard, our own barren savannah. Every day, we burn fossil fuels for everything from flying jets to heating our houses, and in the process, we release carbon dioxide into the atmosphere. There, it's building up and wrapping around the planet like a thick, fluffy blanket, trapping heat inside. And that heat has consequences: Earth's glaciers and ice caps melt; hot air brews into hurricanes so strong they leave cities trembling in their wake; droughts strip moisture from once-fertile farmland.

Just as the lions blame the hyenas in *The Lion King*, Americans spend a lot of time pointing to overpopulation in low-income countries and energy use in China and India as if those are the only causes of climate change. But the truth is that just one hundred corporations—mostly oil, gas, and coal companies—are responsible for some 71 percent of greenhouse gas emissions. They're getting rich while the world burns. And generally speaking, each of us in the United States is responsible for far more carbon dioxide emissions than almost any other person on the planet. In 2014, the average American released more than twice as much carbon dioxide into the atmosphere than the average Chinese citizen and ten times as much as the average Indian citizen. From 1850 to 2014, our nation sent more carbon dioxide into the atmosphere than any other.

Even though we know all this, our representatives dawdle on climate reforms, if they even believe in climate change in the first place. We keep driving, flying, and heating and cooling our houses with oil and gas. I keep doing it even though I know the science. Tackling a problem so huge, so communal, seems impossible.

While we dawdle, our warmer atmosphere heats the ocean, and that balmy water may affect the ability of krill to grow and reproduce, and krill feed creatures from small fish all the way up to blue whales. Without krill, the whales, and so much else, will starve. A warmer ocean can also harbor new diseases and parasites, and will carry sound differently, making it harder for whales and other animals to communicate. While the harm we're causing whales may, in an unsettling move, cause a brief uptick in whale falls, if populations plummet, so too will the habitat their bodies provide.

If you take something out of Earth's natural system, like the carbon that's been buried deep within the ground as coal and oil, the system will wobble until it finds a new and different balance. We just might not like the balance it ends up with.

We've watched this story play out time and time again; we even include it in our children's tales, and yet we don't ever seem to learn. Eat all the wildebeest, drink all the water, and you'll be left with nothing.

———

Nearly three million whales were killed by the whaling industry between 1900 and 1999, slaughtered for their oil and bones. That doesn't even factor in the whaling that took place earlier, at the height of industrial whaling, the era of *Moby-Dick* and sub-Arctic expeditions. While it's hard to say for sure, genetic analysis suggests that large whale populations may have been depleted by 85 percent or more by human whaling. Blue whales in particular, those giants of the sea, have been reduced to a single percentage of their historical population in the Southern Hemisphere, their numbers dropping from 327 thousand in 1904 to just a couple thousand in the early part of this century.

At the end of their lives, whales are huge carbon sinks. They eat vast quantities of krill and small fish and transport the carbon in their bodies to the ocean floor when they die. Fewer whales means less carbon capture. Thanks to whaling, literally millions fewer tons of carbon are captured each year.

And if whale falls have historically been oases in a realm of otherwise scarce habitats, these havens have become fewer and farther between since humans started roaming the ocean in large numbers, slaughtering whatever we found. Though many formerly hunted whale populations have rebounded since the international moratorium on commercial whaling in the 1980s, whale populations—and whale falls—are now only a shadow of what they once were. Every year, whales are dying because of us.

The math is simple: fewer whales, fewer whale falls. And fewer whale falls means less habitat overall. Whale fall expert Craig Smith estimates that 15 percent of whale-fall specialist species could go extinct in the near future; in the North Atlantic, a third of the local specialists may already be gone forever.

When we were killing whales, we thought they were endlessly bountiful; we thought we deserved every bit of whale we took. But we weren't just wiping out the whales, we were wiping out everything that depended on them. We may have signed the death warrant for whale-fall specialists before we even knew they existed.

———

For a long time, Europeans believed that whales were monsters. Cetacea, the group of animals that includes whales and dolphins, gets its name from the Greek *kētos,* meaning large fish or sea monster. On old maps, whales are grotesque. Several appear on Olaus Magnus's Carta Marina, one of the most famous illustrated maps of the early Renaissance. On one part of the map, a tusked, green *orcha* attacks a fanged *balena,* both with demonic glints in their eyes. In his map key, Magnus declares that a *balena* may be four acres large, with eyes as big as fifteen men.

Elsewhere on the map, north of the Faroe Islands, a pig-like, turtle-backed whale rises to the surface, so large that men have mistaken it for an island. The men have anchored their ship to the whale and set up camp on top of it, building fires to cook their food. But inevitably, Magnus writes, such whales eventually dive, and the men upon their back, "unless they can save themselves by ropes thrown forth of the ship, are drown'd."

In the early days of European oceangoing, whales were monstrous, queer creatures that sailors only glimpsed in brief moments at the surface. They were so unlike us, so alien, that the only option was to make up stories to try to explain them. As our relationship with the ocean changed, our understanding of whales evolved, though not necessarily for the better: we saw them as risks likely to destroy a ship or eat our catch.

It wasn't until commercial whaling that we began to understand whales more fully. Finally, we could get a closer look; naturalists could examine a whale that hadn't been rotting on a shoreline for days, bloating and sloughing off its features as it decayed. But even then, we saw them as little more than resources. Not monsters, maybe, but still not worthy of our respect. Even now, we're just beginning to understand how intelligent whales are; how they have cultures of their own; how our actions have annihilated their lineages and their communities.

In places where people have lived alongside whales for generations, this relationship looks markedly different. Many Indigenous cultures that have depended on whales for hundreds, if not thousands, of years—the Iñupiat of what is now called Alaska, the Makah and Nuu-chah-nulth of the Pacific Northwest, the Yupik in the Bering Sea—have long ascribed agency, culture, and language to whales. In Indigenous whaling communities, hunters understand that the whales have their own ethos and that they can choose whether to give themselves up, depending on how the hunters behave. To those groups, whales are not monsters but equals. Even if whales aren't human, these cultural practices affirm that they deserve our respect.

———

My earliest memory of the ocean is *The Little Mermaid*. Like *The Lion King*, *The Little Mermaid* held a powerful place in my childhood. I owned stuffed animals of Flounder, Ariel's best friend, and Sebastian, her reluctant guardian crab, and took them with me everywhere. I watched the movie, which had been released the same year I was born, repeatedly on VHS. It was my first entry point to the ocean, and Ariel was one of the first role models I ever had of a girl who was independent, spunky, and interested in getting to know the world beyond her home. Even as a teenager years later, the movie held such a nostalgic place in my heart that when my mom made flounder for dinner one night, I couldn't bring myself to touch it.

But then there was Ursula, the sea witch, who would look at home on the Carta Marina. At the end of the movie, she conjures a vast whirlpool and grows massive and imposing, a veritable kraken from the depths, and I quaked in my seat every time I watched it. One of the first nightmares I remember having involved a menacing octopus, no doubt inspired by Ursula's disturbingly flowing black and purple octopus legs.

Ursula is paired with still more monsters, her pet eels, Flotsam and Jetsam. They lurk in her sea cave, sinister, ready to do her bidding. Together, the eels slither through the sea, their toothy grins and mismatched eye colors hinting at a certain untrustworthiness, a willingness to bite. It is Flotsam and Jetsam who convince Ariel to go to the sea witch for help becoming human so she can be with the man of her dreams. Thanks to the eels' wily encouragement, Ariel becomes a pawn in Ursula's quest for power, signing away her voice and her soul in the process.

But now when I look at Ursula, it's hard not to be enthralled. She's powerful, self-assured, and delightfully weird. Ariel, on the other hand, makes me roll my eyes. Sure, she's a scientist of sorts, seeking artifacts and answers about a world she can't access, and I respect that, even if her sources are dubious. But she also gives up her voice—the very thing that brings her the most joy—for a man she's never even spoken to. Watching her, I want to reach back in time to my young self and remind her that what seems at first different and frightening is often the best of all.

———

In the original "The Little Mermaid," the fairy tale by Hans Christian Andersen, the mermaid never gets her prince. Rather, she is an allegory for unrequited—and queer—love.

Andersen was bisexual, and in 1836 he was deeply in love with his friend Edvard Collin. "I long for you, yes, this moment I long for you as if you were a lovely girl," he wrote in one letter to Collin, adding that no one had ever "been loved so much by me as you." But alas, their love was not to be: Collin did not return the affection. Andersen wrote "The Little Mermaid" while avoiding Collin's wedding, channeling his queerness through the mermaid's aberrant body and his frustrated love in her failure to capture the prince's affections.

In both the movie and the fairy tale, the mermaid must marry the prince to survive. But in the fairy tale, the prince marries another woman, unwittingly sentencing the mermaid to die. At the wedding, the mermaid dances "gloriously" to please her prince but with a sharp pain in her heart: "She knew that this was the last evening she would see the one for whom she had abandoned her family and her home, given up her lovely voice, and suffered endless daily torments, although he never knew." As Andersen had to suffer, so did the mermaid—two queer souls all alone.

Disney, of course, left out any hint of gay allegory in *The Little Mermaid*. Ariel defeats Ursula, gets the guy, and presumably lives a long straight life on land—long enough, at least, that there's a *Little Mermaid II*.

With its legion of princesses who fall helplessly in love with milquetoast princes, Disney is a veritable bastion of compulsory heterosexuality. But there is queerness if you know where to look: gay-coded Disney characters have been around for a long, long time. Alongside Ursula, there's *The Lion King*'s Scar, *Hercules*'s Hades, *Pocahontas*'s Governor Ratcliff, *Aladdin*'s Jafar. They're almost all villains, obsessed with power, preying on those less powerful than them, and also queeny, flamboyant, and concerned with their appearance.

Ursula herself is based on Divine, the legendary drag queen: gay lyricist Howard Ashman built the character after an early sketch reminded him of John Waters's muse. Ursula's eye makeup reaches up to her hairline like Divine's campy eyebrows in *Pink Flamingos*; her body, half octopus, is unlike any of the other mermaids'. Her voice is deep and thick, tinged with androgyny. Just a few years later, Scar would become her slinky gay counterpart in *The Lion King*, effeminate from his first moment on screen. There's an impeccable moment in "Be Prepared," his manifesto song in which he plots Mufasa's death, when he essentially struts down a catwalk and flounces his mane in the true spirit of haireography.

Their queerness puts them outside the natural order of things, creating an obvious wickedness. Ursula subverts feminine beauty standards and forges power from her curviness and fatness. She urges Ariel to rely on sexuality to get the guy and simultaneously rejects and covets the power held by the chiseled, ultra-masculine king of the ocean, Triton.

Like Ursula, Scar doesn't care about the traditional transfer of power. He has no romantic partner or apparent interest in reproduction, and children (or at least Simba) seem to downright irk him.

Most problematic of all, these movies seem to posit, is that Ursula and Scar ally themselves with those on the fringes—creepy-looking, androgynous eels and hyenas. They're not like the others, and that simply won't do.

In the original "The Little Mermaid," queerness is a lack, a longing, an impossibility—the mermaid yearns for the prince just as Andersen yearned for Collin, but neither could ever attain their love. But for Ursula, Scar, and the other queer villains of Disney, queerness is power, and even more than that, queerness is *fun*. Their difference gives them drive; it lets them imagine untold possibilities.

One of my favorite things about being queer is that it taught me from a young age that things are not always what they seem. When you've been told your whole childhood that growing up looks like finding a good man, getting married, and starting a family—when Disney movies have told you that you should give up your voice and your essential self to find a dreamy prince—realizing that you'd much rather jump into bed with another woman makes you wonder what else is up for debate.

The logic looks something like this: If my partner isn't going to be the sort of person the world told me they'd be, maybe my family could be something different too. And if my family can look like a group of people who have no reason to care for each other—no blood relation or legal tie—other than the fact that they like each other, maybe the society and economic system that depend on repeated units of a husband and wife and two-point-five children are unstable too. Maybe our world doesn't have to look the way it looks.

Which means, maybe, it isn't just our social systems that are up for debate. Maybe we could think of our very planet and the other organisms that live here differently too. A whale might not be a monster; its decaying body might be a home. We might then recognize that humanity isn't the only part of this world worth preserving.

———

Bowhead whales live further north than any other whale species, spending much of their lives beneath pack ice and using their enormous heads to break through when they need to breathe. They prey on blooms of krill and copepods, sifting these tiny creatures through long plates of baleen and slurping them down. From this feast, they manufacture vast sheaths of blubber, up to a foot and a half thick, to keep them warm in the Arctic chill.

Unfortunately for them, their Arctic blubber also makes them a target. While Indigenous Peoples have been hunting bowheads sustainably for more than a millennium, Europeans only got in on the action in the sixteenth century. For four hundred years, European and American whaling crews hunted bowheads in huge numbers for their meat, oil, and baleen until the whales were almost extinct.

Bowheads have another claim to fame: they are the longest-living mammal on the planet. We first learned that not through genetic analysis or some other high-tech method but because scientists have found stone harpoon points and other whaling technologies embedded in the blubber of modern whales. Harpoons of that sort haven't been used since the mid-1800s. We know that bowheads can live for some two hundred years because we shot at them, and some of them survived.

Put another way: some bowheads have lived through six or more generations of humans.

Some populations of bowhead whales have begun to recover since we stopped commercial whaling. In a sense, in recent decades, they've become some of the luckiest whales—living in the Arctic, they're generally far away from noise pollution, oil spills, and other threats. But that is rapidly changing. As the Arctic warms, it's becoming a shipping corridor, and nations and corporations are eyeing possibilities for oil and gas development, tourism, and other exploitation. Bowhead whales of the future may have to contend with cargo vessels plying their waters, sonic booms deafening them, toxicants poisoning their surroundings, and food sources vanishing after centuries of consistency.

A lot can happen in a bowhead whale's lifetime—two hundred years is an awfully long time. There may be a whale out there who narrowly escaped a whaling harpoon, watched its friends being killed, raised calves, built a community of other whales, and made it to old age, only to watch its feeding grounds and way of life begin to disappear as the polar ice recedes.

2

ICE

Turn the world's clocks back about twenty-five thousand years, and you'll hit our last glacial maximum—the last time Earth was really, really cold. Our planet's orbit and position around the sun change over time: over the course of thousands of years, Earth wobbles on its axis, and the shape of its path alternates between round and elliptical. Depending on where we are in these cycles, we receive more or less heat from the sun. Over the past several million years, our planet has teetered in and out of cold periods, with sea ice and glaciers ebbing and growing as we approach or retreat from the sun.

At that most recent glacial maximum, ice stretched south from the Arctic and north from the Antarctic, grasping toward the equator with greedy hands. In North America, two enormous ice sheets, the Cordilleran and the Laurentide, stretched through what would one day be Canada and the northern United States, coating once-green lands in a white glaze. They formed a cap on the land more than a mile thick in places. The Laurentide took the eastern half of the continent, while the

Cordilleran extended frigid fingers into parts of what are now Washington, Montana, and Idaho, not far from where I live now in Portland. It was a time of deep freeze, of woolly mammoths and mastodons.

Ice is a solid, but when enough of it accumulates, it acts like a liquid. At a glacier's mountain altitudes, snow and ice build up over seasons and years. Their weight bears down, creating so much pressure that older ice beneath them has no choice but to yield and flow downslope. Every time snow falls, a glacier gathers more strength, pushing and growing, pushing and growing.

Though it moves slowly, glacial ice can carry entire mountains forth, depositing huge boulders at its feet and transforming the landscape. Look at much of the North American continent today, its mountains and its valleys, its rises and dips, and you're seeing evidence of former glaciers, water leaving its print upon the rock.

―――――

Ice holds tight to relics from our past, preserving them from the decay of time. In recent years, melting ice has revealed two-thousand-year-old forests, six-thousand-year-old collections of Viking artifacts, ten-thousand-year-old hunting weapons, thirty-thousand-year-old seeds cached by an Ice Age squirrel, and more. Ice can preserve people too: in 1991, hikers in the Ötztal Alps found a mummified body dating back to the Copper Age.

More recently, pop culture has given us another offering from the ice: Captain America. *Captain America: The First Avenger,* the movie that introduced this decades-old character to the non-comic-book-reading masses, begins with soldiers searching an ice-bound ship at the North Pole and finding a shield, red, white, and blue and emblazoned with a star, frozen in a block of ice. Beneath the shield is Captain America, preserved since sacrificing himself in World War II.

I never heard of Captain America growing up. There were no comic book stores in my neighborhood, and the only superheroes I knew much about were Batman and Superman, beefed-up men who saved damsels in distress and then stared, square-jawed and heroic, off into the middle distance. Every once in a while, I watched X-Men cartoons when my brothers had them on, but never with any more consistency than was necessary to develop a wistful crush on Storm. Superheroes just weren't really my thing.

But when I was twenty-five, I encountered the Avengers.

I was living with my then-partner, having followed them from Seattle to New Haven despite knowing that our relationship likely should have ended years earlier. I was miserable, just out of grad school and trying to figure out what I wanted to do with my life and whether I should stay. I was friendless, not sure how to begin to meet people in that cloistered, frigid New England city. I became numb to my relationship and to the world; I curled in upon myself.

That winter, it seemed to snow endlessly. Week after week, I shoveled our front steps and dug out the car before retreating inside from the biting cold and burying myself under blankets. One night, while my partner was at the library studying, I went to Netflix and figured, *The Avengers*, sure, why not? Everyone was talking about it on the internet. I didn't have anything better to do.

It was a stupid movie. I had no idea what was going on or who the characters were. *The Avengers* is an ensemble movie, the sixth in the linked films of the Marvel Cinematic Universe. It depends on you already caring about Iron Man and Captain America and all the rest of those muscled men; it assumes you know who they are and why they act the way they do, why Iron Man and Cap seem to hate each other and what the hell a Norse god is doing in a superhero movie. But it also wasn't just a bunch of jocks running around making sexist jokes. It was funny; it was easy. So I didn't turn it off, and then I watched it again, curling up on the couch and finally beginning to thaw. Then I went back and watched *Captain America: The First Avenger*, which travels back in time after its opening scene to tell the origin story of Steve Rogers and his transformation from scrawny Brooklynite to swole symbol of the nation.

To this day I can't quite put my finger on why I love *The First Avenger* so. It's jingoistic, driven by patriotism and male duty and militarism. What Steve wants most is to enlist in the army—he tries at least five times and is turned away each time due to his various disabilities—so he can serve his country and fight the Nazis. While I relate to his stubbornness, his path is foreign to me. Militaries and war have always struck me as emblematic of the worst of toxic masculinity—nations' attempts to one-up one another with pain and murder and destruction. While I realize I am in many ways privileged to live in the United States, I shy away from our patriotic attempts to assert ourselves as better than other nations, as more deserving of our wealth and power. I should hate Captain America.

But in spite of myself, I loved Steve Rogers as the underdog, the subverter. In his original body, he's hardly the paragon of masculinity: He is scrawny, asthmatic, chronically ill, hard of hearing, and a mere 110 pounds. He constantly picks fights he can't win because he wants to stand up for others, not because he's drawn toward violence. "I don't want to kill anyone," he says when asked if he wants to go to war to kill Nazis. "I don't like bullies. I don't care where they're from." His compassion for people facing injustice is what makes Doctor Erskine, the creator of the Super-Soldier Serum, choose him as its recipient. Steve is a rejection of the toxicity of war; he is an alternate version of masculinity.

I sank into the Avengers, pulled the movies around me like a heavy blanket that could keep out the New Haven chill and help me imagine a different life. No matter what, no matter how bad things got, they would be there for me, ready to save the day.

———

Now, years later, after leaving New Haven and that long-expired relationship, moving to DC for a brief stint, meeting my now-spouse Leslie, and heading west with them to Portland, I live in the shadow of glaciers. Mount Hood is adorned by twelve glaciers, the snowfields visible from the city on a clear day even at the height of summer. Mount Adams, further from Portland, also has twelve, while Mount St. Helens holds the world's youngest glacier, resting in the crater left by the volcano's explosion in 1980.

I'm a hiker, not a mountaineer, so I've never trodden across the ice fields—but I've spent plenty of time on trails near them, with goosebumps on my skin from the rush of wind blasted across the ice. Since we moved to Portland several years ago, Leslie and I have spent most weekends in the mountains, hiking the base and flanks of Mount Hood and other mountains throughout the Oregon Cascades.

The day after we got married, we hiked to Cooper Spur, a promontory high on the side of Mount Hood. Our wedding had been barely more than an elopement: we got married at the courthouse by a delightfully lesbian judge, in a ceremony that lasted five minutes at most, witnessed by two friends. The hike, a journey into wilderness where we were together with the landscape of our home, felt like a more apt celebration of our partnership than any ceremony ever could.

The trail began deceptively calmly, a gentle rise up the mountain. We meandered through windswept hemlocks and across spring-fed meadows, pausing to inhale the verdant cedar scent of alpine growth, to listen to birds flitting and chattering among the trees. Then the trail spat us out of the forest above the tree line, onto switchbacks across a broad slope covered in wildflowers that hugged the ground and expanses of ashy, sandy dirt. Halfway up, the jagged peak of Mount Hood came into view, beckoning.

From there, our route rose brutally, a cindery path through a huge field of boulders. We were following a ridge beside Elliot Glacier, the mountain's largest glacier by volume. It emerges from the mountainside in white reptilian scales, its crevasses the kind of luminous blue you only ever see in ice. It was late summer, and dust covered the August melt. In winter, the glacier would have been shiny, blinding white with new snow and ice.

The view pulled us upward as we slipped and slid. The wind picked up, cooling our skin even as the summer sun warmed it. We swore at the trail and panted and sweated until finally, impossibly, we made it.

From the spur, it's nearly another three thousand feet in elevation gain to the mountain's peak, but standing up there in the thin, crisp mountain air, it seems almost possible to reach the top just by raising your hand and grasping.

It was the highest on the mountain we'd ever been, so high that we were huffing and puffing in the thin air, gasping for a full breath. We were elated, here on this mountain we call home, together in a partnership that felt truly unbreakable.

We'd both felt conflicted about getting married, worried we were surrendering to the patriarchy, doing exactly what straight people did and upholding a system in which only married people got things like health insurance and tax benefits. But we felt, too, like a legal bond might help us steel ourselves against the onslaught that we feared was coming in the Trump administration—that our stability might help us be a safe haven for each other, and maybe for other queers too. Here on this ridge, we felt grounded, hopeful; here, a massive glacier had literally moved a mountain. Maybe our partnership could do the same.

———

In comics and in the movies, Steve is forever shoulder-to-shoulder with Bucky Barnes. In the Marvel Cinematic Universe, their friendship has endured since childhood, Bucky constantly looking out for tiny, scrappy

Steve and getting him out of trouble. When pre-serum Steve stands up to a man heckling the war effort in a movie theater and gets beat up in the alley behind the building, Bucky is the one who pulls the bully off him. And Bucky is Steve's sole support as his mother is dying. "I can get by on my own," protests Steve in a flashback to their early adulthood, but Bucky knows that isn't exactly true. "The thing is, you don't have to," Bucky replies. "I'm with you to the end of the line."

And Steve follows Bucky with just as much dedication, playing wingman on double dates and keeping Bucky from getting too cocky. When Bucky gets his orders to go to the front, pain and worry and jealousy all flash across Steve's face. Then Bucky is gone, leaving Steve alone on his journey to superhero status.

That is, until Steve—now Captain America—winds up in Europe and hears that Bucky's division has been taken captive. Steve disobeys orders, commandeers a plane, and drops behind enemy lines alone to save his best friend—and succeeds. In the Nazi stronghold, Steve pulls a weakened Bucky off a sinister operating table, and they fight their way out of the burning building. Together, they become the heart and soul of the Howling Commandos, an elite squad Cap leads to take down the Nazis' rogue science division, Hydra.

On a mission, Steve loses Bucky once more: A Hydra soldier blasts a hole in the train they've infiltrated, leaving Bucky dangling from the side over a deep ravine. Losing his grip seconds before Steve can save him, Bucky plummets hundreds of feet into an icy river.

Bucky's apparent demise leaves him intertwined, like Steve, with snow and ice. Though assumed dead, he survives—the operating table Steve rescued him from was, in fact, the site of his initial transformation into the world's second super-soldier. But his fall into the frozen abyss lands him back in Hydra's hands. Hydra operatives drag him bleeding through the snow and into surgery, and Bucky continues his hibernal transformation.

While Steve slumbers in a frozen plane somewhere around the Arctic Circle, Bucky is turned into a weapon. Hydra regularly wipes his memory and conditions him to follow the orders of anyone who uses a key sequence of words. No matter how hard he tries to fight, once triggered, Bucky becomes the Winter Soldier, an assassin ready to comply. And whenever he's not needed, Hydra puts him in cryostasis for months or years at a time.

For Steve, ice represents lost time, but for Bucky, much more is lost beneath the ice. Memory, control—and home.

———

My home in Portland bears the fingerprint of vast ice fields. Fifteen thousand years or so ago, the Cordilleran ice sheet shifted and grew, sending out a frigid tentacle across what is now the Clark Fork River in Idaho. Stuck behind the ice dam, the water pooled into an enormous lake, Lake Missoula, twice the size of the Great Salt Lake today and nearly half a mile deep. Where today's lakes are bound by soil and rocky banks, Lake Missoula's shores were made of ice, fragile and fluid in a way earthen stone and soil are not. Every so often, the lake would fill to the brim with glacial melt, and the pressure would build and build. Water would force its way into crevasses until finally, the ice dam would give way and the lake would spill free, a vast flood known as a *jökulhlaup*. Then the glacier would extend again and create the lake anew.

It was a cycle that happened maybe forty times over two thousand years: the lake would fill like a bathtub about to overflow, the ice would float away from the lakebed, the dam would collapse, and water would surge into the surrounding land. The water escaped the lake at nearly sixty miles an hour, carrying icebergs and tearing through the forests of Montana and Idaho, picking up dirt and gravel and even boulders the size of cars, and stripping eastern Washington down to its bedrock.

Today, if you drive east out of Seattle, past Washington's mountains and the farms nestled into the foothills, you'll reach the Channeled Scablands, evidence of the many Missoula floods. Canyons crisscross arid plateaus, forming a maze of rocks and dry channels. It looks as if some ancient knife has slashed its way across the land. Hills roll across the scablands, giant ripple marks left by the rivers that rushed out when the ice dams burst. Where water still flows, cataracts gush down narrow gorges. And atop lonely buttes, glacial erratics stand sentinel, huge boulders carried all the way from Lake Missoula on the backs of icebergs amidst the silty floods.

The water didn't stop in eastern Washington. Once it took all it could from the scablands, it kept going, spurred on to the floodplains of Washington and Oregon by momentum and gravity, intent on making its way to the Pacific Ocean. The deluge scraped the sides from the Columbia River Gorge, ripping Douglas firs and hemlocks from their roots and carrying them out to sea.

In the biggest floods through the gorge, the waters reached more than 700 feet above where the river flows today. It brought violent, rapid change to the landscape: while some of our continent's largest canyons, like the Grand Canyon and the Black Canyon of the Gunnison, formed over millions of years, the powerful Missoula Floods eroded most of the Columbia River Gorge in just a couple thousand.

———

Imagine living along the river, like I do, and knowing that perhaps twice in a lifetime, the ground would rumble and the floods would come, rupturing the landscape. I don't know how I would cope with that inevitability, the knowledge that everything I've built and gotten used to might soon be stripped away by a cascade of muddy water.

But we're heading toward a time when I might *need* to start getting used to it, or something like it—when we all might need to. Glaciers on every single continent on this planet have retreated over the last century; some are gone completely. In Alaska, the Muir Glacier—which John Muir himself described as a "spacious, prairie-like glacier"—retreated a massive thirty-one miles between 1892 and 2005, a distance more than twice the length of Manhattan. It's not the only one to experience such diminishment.

Some 10 percent of our planet is covered in glaciers. If you put all the glacial ice on Earth in one place, it would cover the United States and Mexico and jut into the lower reaches of Canada. But these ice sheets are receding, and as they melt, they'll turn to water, rushing down rivers and scouring out the land.

The glacial meltwater eventually reaches the ocean, which could, as a result, rise by up to eight feet by the end of the century. That's far more than it sounds like—eight more feet of ocean would swamp almost every major coastal city, turning streets into canals. Already, Miami experiences what's called sunny-day flooding, when high tides inundate low-lying streets. Louisiana's coastline is rapidly eroding. On islands throughout the world—the Solomon and Marshall Islands, Palau, Samoa, Tuvalu, and others—rising saltwater is mixing into groundwater sources, contaminating what little drinking water is available.

And if we do somehow manage to melt every chunk of ice there is to melt, as the ocean warms, the water within it will expand and continue to rise. Once the glaciers are all gone, our world might just keep drowning, leaving us with no choice but to flee to higher ground.

It seems impossible to think that we might one day just simply not have our ice. No frozen cap on the North Pole, no frosted landscape in the Antarctic. But it's happened before: some eighty-five million years ago, it was warm enough in Antarctica that flowers and trees thrived; fifty-two million years ago, alligators lived in the Arctic. Even though it's unlikely that we'll reach quite that level of transformation, at least not in the next few centuries, at the rate we're going, we'll soon be living in a very different world—and we're still learning just what the differences will be.

For one thing, sea level rise is deceptively difficult to predict. If the ocean rises by a foot, we think we should be able to draw a line around all the places that are a foot above sea level today and know exactly what's going to be underwater in the future. But in reality, the seas won't likely rise so evenly.

When our ice sheets melt and thin, their weight will vanish from the land they once sat on. In their absence, the land in those places will rebound, springing subtly higher. So even as the seas rise, the land in some places will too: places like Greenland and Canada, today coated in ice, will likely feel the effects of sea level rise less as the land shakes off its icy weight.

Other places near the tropics don't have that luxury. That land will stay at the level it's been, even after the ice melts. But all the water that was once ice will need to go somewhere, and it's likely to wash its way up their shores and into their freshwater sources.

For centuries, Europeans and Americans have exploited people throughout the tropics. We colonized everywhere from Bangladesh to Indonesia to small islands in the Pacific; some of these places, like American Samoa and Guam, are still essentially colonies. We've damaged those lands with cash crops, mining, and weapons. The Marshall Islands are still contaminated from nuclear testing; Nauru's environment has been ravaged by decades of phosphate mining. Much of Hawaii's indigenous flora is gone forever thanks to ranching and farming, and rain erodes the islands and smothers the fringing coral reefs. The list goes on and on.

Now, climate change is threatening these places with sea level rise, ferocious hurricanes, and more. The people of these tropical nations and regions aren't the world's great emitters of carbon; they aren't the ones who created this mess. But the people capitalism has so often penalized are harmed once again.

———

For the last 11,500 years or so, we've been in what's known as an interglacial period. The ice age of our ancestors technically never ended, even though the giant ice sheets have vanished. *Ice age* just describes any time our planet has ice at its poles and in the mountains, and that's been the case for the last two and a half million years. But at this very moment, we do have a lot less ice than we used to. We're in an ebb, a liminal space between tons of ice and none at all.

I've always resided in the in-between. For a long time, it made me lonely. As a teenager, I didn't fit into one friend group—I didn't have the words for it then, but I felt too queer for my straight friends yet too nerdy and too straight-edge for the artsy queer kids I knew. I felt a constant thrum of anxiety in my chest, an ache of sadness that weighed me down and made it hard for me to look people in the eye.

As I've gotten older, I've learned to embrace my liminal self, to find satisfaction in feeling not quite right. It keeps me from feeling too settled or used to the world; it opens me to possibility. Through my queerness, I've found people who embrace the same unsettling, in my friendships and in my relationships. That common language of in-between—of fluid sexuality and gender—makes sense to me.

I like what theorist José Esteban Muñoz has to say about queerness. "Queerness is not yet here," he writes. "Queerness is an ideality. Put another way, we are not yet queer. We may never touch queerness, but we can feel it as the warm illumination of a horizon imbued with potentiality." Queerness is an act of *becoming*, a looking toward a utopia, but an ambiguous one, full of possibilities.

Liminality drew me to Steve Rogers: a man out of time, he can't go back to the forties but also doesn't fit quite right in the present day. It gives him a perspective many of the rest of us often lack.

In *The Winter Soldier*, he learns that S.H.I.E.L.D., the world's major defense organization, is building a superweapon that can target "threats before they even happen." Just as queerness's dislocation from the heterosexual mainstream lets us imagine another society, a utopia, Steve's jump through time gives him enough distance to see the slow creep of modern surveillance technology for the threat it is. Steve isn't a super-soldier because he wants to support weapons and death, but to protect people. "This isn't freedom, this is fear," he says. There is another way.

What would the world be like if instead of focusing on conquest, the United States focused on preparing for—or even slowing—the effects of climate change? What if we focused on people, not capital; on our communities, not ourselves? If we put even just a fraction of the money we spend on our military into climate technology, think of the emissions we could curb, the carbon dioxide we could draw out of the atmosphere. Think of the money we could spend on making sure people have a warm, dry bed—one not swamped by sunny-day flooding—and food not scorched away by increasingly frequent summer heatwaves and wildfires.

Right now, our planet is at a turning point within a turning point, a crossroads within the interglacial. We have a choice of what we will become, but I'm not sure we're prepared to make it.

———

In college, I had acquaintances who were focused on climate change. They studied environmental and political science, and on weekends they journeyed from our rural enclave to cities where they protested government inaction and met with other students to plan how they would change the world for the better.

I spent a lot of my time in nature then, and I respected the work they were doing, but I didn't feel it was for me. Climate change didn't seem urgent. My time would be better spent, I thought, on working for queer rights and social justice. The nature I knew wasn't political.

But over years, an awareness of the immensity of climate change accrued in me like ice and snow piling up to become a glacier. I became friends with people studying the environment and how to fix it; I picked up a freelance gig writing about the drought in California; I started reading about the climate, first a few articles here and there, then book after book. Climate change became real; it became pressing.

I don't remember when I realized that climate change isn't just about the environment but also about society, but I know it changed the way I think about the world. Queer rights, racial and gender equity, ending poverty—none of that matters if our world is underwater, with us gasping at the surface for air. And those who are most vulnerable in a climate-steady world—people without homes, people without the money or the social capital to just pick up and move if they need to— are the ones who are impacted first as our world becomes less livable. If I care about social justice, I have no choice but to care about climate change.

I'm not the only one who has started to care: according to Yale University's ongoing research on Americans' opinions on climate change, the number of people who are "alarmed" about climate change more than doubled in size between 2015 and 2020.

Even if most of us haven't been focused on it, climate change has been happening for a long time. The first paper about the harmful effects of carbon dioxide was published in 1896. That paper pinpointed the Industrial Revolution as the beginning of massive changes to our planet, and researchers today tend to agree. That's when we first started pumping fossil fuels into our atmosphere, when our planet started warming at an unprecedented rate.

We've been making these changes for generations and generations, but most of us are only now beginning to notice what we've done.

———

Captain America keeps a list of all the things he's missed during his time in the ice, a shorthand for all the ways the world can change in just a few generations. On the list are *I Love Lucy*, the Moon Landing, the Berlin Wall, Steve Jobs and Apple, disco, Thai food, *Star Wars* and *Star Trek*, Nirvana, *Rocky*, and the *Trouble Man* soundtrack.

Missing from that list are transformative moments like the Civil Rights Movement, Stonewall, the Vietnam War, the War on Drugs. There's no Cold War, no Red Scare, no *Roe v. Wade*, no war in Iraq. There's the Berlin Wall, yes, and the Moon Landing could be read as a gesture toward the US–Soviet relationship, and the *Trouble Man* soundtrack an even more oblique reference to civil rights. But for the most part, it's an easy list, one that touts consumerism in place of the complexities of social history. It's primarily a joke—in each country the movie aired, the list is a little different—and the humor distracts from just how overwhelming it would be to skip ahead to the future.

That tidiness is part of what made Marvel movies so compelling to me when I was depressed and lost. In each movie, a horrible attack overwhelms the normal capacities of humans to defend ourselves, and superheroes save the city or the world. In *The First Avenger*, Cap takes on an evil army fueled by the Tesseract, a mysterious, otherworldly power source that can be used to create unstoppable weapons. In *The Avengers*, aliens invade New York, instantly overwhelming the New York Police Department, and the Avengers vanquish them. In *The Winter Soldier*, Cap and his companions quell an attempt to wipe out millions of people. Time and time again, our heroes face unreasonable odds and come out on top. The aliens are killed; humanity is saved.

Movies help us hold a mirror up to the world and understand how we see ourselves and the societies we live in. One of the reasons Marvel movies are so popular is that they take a chaotic world under threat and make it safe, all thanks to one superpowered group of people.

But movies also help us escape. In the air-conditioned darkness of a movie theater, we can pretend our world isn't warming at a terrifying pace. We can forget that we're causing a mass extinction of everything from insects to whales. We can tell ourselves the latest balmy winter or wretchedly hot summer was just a fluke.

It's so easy to believe someone else will take care of things. I'm actively worried about climate change, yet I assume someone else will invent the world of the future I've been promised, where all our vehicles are electric, our windows generate clean power, we no longer have to extract from the earth, and our air is sparkling clean. Someone else will figure this out eventually, right? People are working on this.

There are nearly eight billion people on this planet. If we wait for the Avengers to show up, we're in for a world of disappointment. If we don't all show up every single day to deal with this problem, then frankly, we're doomed.

———

Steve and Bucky are bound together by ice, but there's something more. I say I love Steve for his persistence, for his kindness, and his ardor, and it's true: I do. And I say I am drawn to superheroes for the comfort they give me in a world on fire. But really, I love Steve and Bucky for the way that they are family. I love them because they are so very, very queer.

Oh, not explicitly—Marvel would *never*. Not Marvel, who so proudly proclaimed there would be a groundbreaking gay moment in *Avengers: Endgame* that turned out to be just an unnamed character who refers to going on a date with another unnamed man in a scene that barely lasts two minutes. Sure, it's entirely subtext, but look at Steve and Bucky's relationship, and it's hard not to see two men utterly devoted to one another.

Time and time again throughout the Marvel Cinematic Universe, Steve risks everything for Bucky. In addition to his rescue mission in *Captain America: The First Avenger*, Steve spends two years after *The Winter Soldier* searching for Bucky once he realizes he's alive. He quits the Avengers in *Captain America: Civil War* when staying would mean imprisoning Bucky for his crimes as the Winter Soldier, resigning himself to a life in exile. When Steve loses Bucky once more in *Avengers: Infinity War*, he drops to his knees, horror and disbelief on his face as he gingerly touches the ground where Bucky once stood.

Bucky, for his part, remains faithfully by Steve's side throughout the movies. He sets aside his jealousy after Steve transforms from skinny kid in need of his help to super-soldier beloved by all. Within the Howling Commandos, he's Steve's right-hand man and perpetually has his back as the group's sniper.

Their bond is what breaks Bucky from Hydra's control. At the end of *The Winter Soldier*, they battle aboard a crashing aircraft, Bucky still brainwashed and unable to recognize Steve. Even as the Winter Soldier pummels him, Steve sees only Bucky. "I'm not going to fight you," he says, dropping the Captain America shield through the disintegrating aircraft into the Potomac River below. The Winter Soldier slams into him, pinning him down and punching him until Steve, nearly unconscious, says only, "Then finish it. Because I'm with you to the end of the line."

In a moment of flashback and recognition, Bucky surfaces from the Winter Soldier and lets Steve go. Steve drops through the air into the river, where it seems he'll drown, but Bucky dives in after him and fishes him out before vanishing into a journey to regain his memories.

There's an argument to be made that theirs is just a uniquely close platonic friendship—but for me, that argument has, as culture writer Kaila Hale-Stern describes it, "strong 'historians say these passionate love letters between two men were just how people used to talk to their friends in olden times' vibes." And I'm hardly the only one who sees it: of the four hundred thousand or so Marvel Cinematic Universe fanfics available on the Archive of Our Own as of February 2021, more than fifty-three thousand focus on the love between Steve and Bucky—making it the fourth-most popular pairing on the site. Though the romantic connection is only implicit, their relationship has a depth and a commitment and a yearning that many of us queers recognize from our own oft-closeted relationships, and a sense of us-against-the-world that rings true for queers living under homophobia. In a time when I—along with so many other queers—was hungry for queer representation on screen, Steve and Bucky's unspoken coupledom helped me feel a little less alone.

————

As our glaciers melt, we're also losing a physical link to our shared history. When snow falls on a glacier, it traps tiny pockets of air among the flakes. These bubbles get preserved as the snow compresses into ice and can remain locked up in this frozen record for thousands and thousands of years. Scientists can read a core drilled out of a glacier the way you might look at rings on a tree, each layer the story of a single year, the air captured within reflecting both global and local climate conditions. The bubbles contain chemical markers that can tell us about air temperature, rain and snowfall, volcanic activity, pollution, and more. They're tiny time machines, letting us glimpse the world as it was before we had instruments to record its conditions, or before we even existed. In Greenland, ice cores can take us back to 123 thousand years ago; the Antarctic ice record goes back eight hundred thousand years through an ice core almost two miles long. To put that in perspective, our species, *Homo sapiens,* evolved about three hundred thousand years ago. Antarctica holds the story of our entire existence and then some.

And thanks to these records, we have glimpses of local and global conditions not only at the poles, but also wherever there's an alpine glacier. Ice core records have been obtained from tropical and subtropical mountains in Tanzania, Bolivia, China, Peru, Indonesia, and elsewhere, allowing us to compare places across different latitudes and climates at similar times in the past.

But meltwater running through a glacier obscures the chemical signals scientists need to understand our past. Or worse, the record vanishes completely as the glacier recedes. On the Quelccaya Ice Cap in Peru, some sixteen hundred years' worth of ice has disappeared. On Kilimanjaro, at least 86 percent of the ice cover that was present in 1912 has completely melted. As the glaciers drip away, we lose opportunities to understand how the climate impacted the cultures that came before our own.

Elsewhere, we've lost more recent records. Mt. Ortles in Italy and Geladaindong in China no longer record the time between 1980 and the present. Naimona'nyi and Nyenchen Tanglha in Tibet have lost their ice back to 1950. The bubbles in newer ice are used as calibrators for the rest of the core, since their composition can be compared to measurements of the atmosphere we've made with modern instruments. Without core samples that represent these recent years, it's far more difficult to reliably understand the samples that show us deep time.

While Steve was frozen beneath the North Pole and Bucky was being cryogenically frozen and thawed, the ice sheets were dwindling away. Just as Steve and Bucky missed out on decades, never to experience them, our collective memory is vanishing into thin air.

———

A few years ago, Leslie and I hiked the Highline Trail in Glacier National Park. It's a long, rolling trail that skirts the Continental Divide. It was early July, the same summer we got married, and the trail hadn't been open for the season very long; in places snow still covered the path, packed and slushy from all the feet treading on it. Where the snow had melted, wildflowers were blooming, beargrass and lupine feeding excited bees that flitted from flower to flower. We began early in the morning and hiked all day, finishing when long shadows were flowing off the mountain tops. The whole way, the views were so stunning it seemed like we had been transported to another planet entirely.

Glacier National Park gets its name from, yes, its glaciers. The staggeringly beautiful mountains have been shaped by bodies of ice carving and pushing stone around over millions of years, leaving their mark for future generations. In 1850, some eighty glaciers glinted like mirrors on the mountaintops, ice sheets that dripped into lakes and rivers and powered an entire alpine ecosystem. By 2015, only twenty-six glaciers were left, and today, there may be even fewer. By 2030, they may all be gone.

We'd driven from Portland to Montana rather than flying, a trip that saved us some carbon emissions—but still, we journeyed the six hundred miles spewing fossil fuels all the way. In coming to see the glaciers, to celebrate and enjoy the beauty of the park, we'd contributed to their melting, and I don't know how to reconcile that.

Over the past few years, I've seen ad campaigns and heard my nature-loving peers talking about going to "see things while they're still here." Go see the glaciers of Glacier National Park before they've dried up; see the Joshua trees before they shrivel and die; see the Great Barrier Reef before the corals bleach; see Venice before it's lost beneath the waves like the city of Atlantis. See the Amazon before we log it all to make room for cattle; the Galapagos before we ruin it with tourism; Alaska before it sinks into a soup of melted permafrost.

I don't want to see these places before they're gone. I don't want them to exist only as part of our collective memory. I want us to save them. But to do that, we all have to look each other in the eye and admit what we've done—what we're doing—and realize that we have to be the ones to force the change. It can't just be some of us, and the change can't just be a tiny shift. We have to be our own heroes.

3

WILDS

Each winter, the Arctic becomes cloaked in a heavy, ghostly darkness. As the northern reaches of Earth tilt away from the sun, vegetation goes to sleep, covered in blankets of snow and crusts of rime ice and hoarfrost. In the infinite night sky, greens and pinks dance and sway above the earth, lighting the path of caribou and hares as they search for a midnight snack. The world is quiet.

Come spring, the sun emerges tentatively from its slumber. At first, it barely shows its face, the only hint of sun the dawn-blue light that settles in the corner of the sky for a few fleeting moments. Then the sun appears for minutes, an hour. The snow and ice begin to melt. The world awakens.

At the height of Arctic summer, the sun transits in great loops above the horizon, never leaving the world in a scrim of total darkness. At most, it dodges horizontally behind a mountain for a few minutes, leaving the world in pale twilight until it emerges on the other side.

Even in the light-filled days and nights of summer, much of the soil remains locked up in permafrost, ever-frozen. Still, the land is lush with tundra, the earth encrusted with mosses and lichens and stippled with low shrubs, like bearberry and willow. Grizzly bears wander between rivers in search of salmon and berries; ptarmigans take shelter among shrubs. Caribou wander vast distances to find the choicest lichens, leaves, and grasses. The tundra cushions their feet as they walk, a thick mat springing gently back beneath them. But even on the balmiest days, dig deep enough and you'll find a sheen of ice crystals. At least, you will for now.

———

When I was fifteen, I spent a month backpacking in Alaska's Northern Talkeetna Mountains. The mountains are below the Arctic Circle, but not by much; it was August and we never saw true night. I was with a group of other teenagers, twelve or so of us, part of a backpacking and wilderness skills course run by the National Outdoor Leadership School.

We came from all over the country, and none of us knew each other before we met up in the small town of Palmer just outside of Anchorage. But all of us liked the outdoors, and all of us had spent at least a little time backpacking. I'd been hiking since I was a kid, and started backpacking as a teenager, participating in backpacking, kayaking, and climbing programs that served as an alternative to more typical summer camps. My first excursions were along parts of the Appalachian Trail near where I grew up. Then as I got older, I went further afield, spending summers in Washington, British Columbia, and Wyoming, lands with bigger mountains than the eroded peaks I knew from home.

But none of that really prepared me for the remoteness and the size of the Alaska mountains. Everywhere else I'd ever backpacked, we'd charted our course along existing networks of trails, stopped at established campsites where others had stayed before us. Here in Alaska, though, we were simply dropped off by the side of the highway in the mountains at a spot that our trip leaders had deemed adequate. We laced up our boots, then looked at our topographic maps and scoped out a potential campsite that was close to water and seemed likely to be flat enough to pitch a tent. Then, we just picked up our packs and started walking: no path, no signs, no single file, just the simplest route we could each find from point A to point B.

As we walked off into the wilds, I was giddy and disoriented. My feet were clumsy over the domed tussocks rising from the ground, and I couldn't take my eyes off the mountains around me. I felt as if I had wandered off into a book, a place more myth than real.

Each morning, we would gather as a group to go over the map and determine where we'd camp that night. Then, we'd split into smaller, more manageable groups, and each group would chart its own course to the X marked on our map. Over the entire month, we never walked a single trail, except for the odd path carved by wildlife through tangled riverside shrubbery.

We were all used to trail networks, to traveling paths people had already trod before us. Our instincts were to walk in a line, to follow the person ahead of us as they found the best route over the springy ground. But walking one after another, we would damage the fragile summer greenery with our heavy, clunky steps. We had to unlearn our hiking habits, to spread out over the tundra, a spray of teenagers across the landscape. It was jarring, requiring me to rewrite almost everything I'd known about backcountry travel—but it was also exhilarating.

While I was there, I defined wilderness in my journal as "the ability to walk for miles and never see another sign of humans." To me, these mountains were the very peak of wilderness. We were in a realm without humans but lush with animal life. We saw moose and marmots, and caribou—so many caribou, herds so big I couldn't begin to count them. My favorites were the juveniles, with their awkward long legs and dinner-plate hooves.

One cloudy afternoon, we rounded a bend in a valley, and a distant herd of caribou, hundreds-strong, came into view. A group of several cows and a bull with a full rack of antlers approached to investigate us, curious. I ached to get close to them, but instead we yelled and made ourselves as big as we could to make it clear we weren't worth messing with—even grazers can be dangerous if they think you're a threat. A few nights later, we camped in a glacial moraine where so many caribou stretched up the snowy slopes that it looked like the mountains were covered in spots. I spent all evening watching as the mothers ate and their babies frolicked awkwardly around them.

We yelled frequently, several times a minute, to warn grizzly bears of our presence so they'd stay away, and always traveled in groups so we'd be safer if we ever did encounter them. It was both a relief and a disappointment that we never came across any, only saw their scat left behind as proof that they existed. Each time I scanned a mountain slope, I secretly hoped that I might get to see the enormous bulk of a bear browsing among the berries—but I knew that it was better for all of us, bears and people, if they stayed away.

Humanity did seep through from time to time. We picked up deflated mylar balloons that had wafted on air currents from warmer climes, and once a week or so, an airplane would fly overhead, both jarring interruptions that left me reluctant to reunite with the rest of humanity. But mostly, we were alone. As far as I was concerned, it was the best, wildest place in the world.

———

I didn't realize it then, but my definition of wilderness was almost quoted verbatim from the Wilderness Act of 1964. The Act defines wilderness as "an area where the earth and its community of life are untrammeled by man, where man himself is a visitor who does not remain." When it was signed into law, it designated some fourteen thousand square miles of land as wilderness.

The Wilderness Act itself comes out of a tradition of American public lands advocacy usually credited to men like John Muir, Ansel Adams, Teddy Roosevelt, Gifford Pinchot, Edward Abbey, and Aldo Leopold. These forefathers of our national parks and forests raised their voices to protect important lands from what they saw as human overreach. John Muir was instrumental in the creation of national parks: he lived and traveled throughout California, particularly Yosemite, and he saw the damage caused to the landscape by sheep ranching and other industries. Thanks to his words and Ansel Adams's photographs, which show the epic grandeur of Half Dome and other landmarks, Yosemite became one of the first national parks in the United States. Not long after, Gifford Pinchot and Teddy Roosevelt would see the wreckage left by widespread logging in the West and establish the Forest Service.

When I'd kayaked in Yellowstone National Park years before I went to Alaska, I could do so because Muir and his contemporaries pushed for an expanded system of parks. When I'd backpacked through the national forests not far from Seattle the year after that, I was walking through lands that Teddy Roosevelt and Gifford Pinchot had fought to protect from rampant logging.

These were the happiest moments of my teenage years, and I owed it all to them. Every time I entered the woods and the mountains, I was walking in the footsteps of these great men who had made sure the forests and the tundra would still be around more than a century later, when a young queer woman would need refuge.

———

I grew up in nature, or at least adjacent to it. Once a month or so, my dad and I would hike a loop along the Potomac River, scrambling across the rocks that rose above the water and watching our dog jump like a mountain goat from crag to crag. As teenagers, my brothers went off to backpacking summer camps, and when I got old enough, I wanted to do the same. I was always trying to keep up with them, determined to do as well as any guy on the trail.

I chose my college largely for its outdoor opportunities too. It was nestled in the Berkshire Mountains in Massachusetts, close enough to trails that a determined person could walk to them from campus. For my campus job, I spent four hours a week taking students on short beginner hikes to fulfill their physical education requirements. Each year on one October Friday, the college president would cancel classes, and the entire school would wander up Mount Greylock's trails together. It was idyllic.

But while I was there, I had coursework to keep up with, classes harder than I'd ever taken before, and then there was the unfathomable task of making friends. I felt alienated from the football and baseball players I lived with and not nearly as smart as my classmates. I struggled to balance the different parts of myself and hated myself for being unable to find a place and people that felt comfortable. Here, I couldn't just slip into the wilderness and forget the rest of the world. Instead, I began to splinter.

I skipped out on outdoor adventures with would-be friends because I was sure I wasn't strong or capable enough. Social and academic worries pulled me away from the mountains, and I was afraid I no longer knew how to be in them. The entire time I was in college, I only went on a single backpacking trip. I got out of shape and stopped hiking with the few friends I'd begun to make after an excruciating snowshoe excursion where I panted, short of breath, the whole way to the top, and spent each step wondering why it was easier for everyone else. I stopped climbing because I worried I would never be as good as my peers. I questioned every single thing I did.

I wanted to be as comfortable in the world as I'd once been hiking along rivers and beneath mountain ridges, but instead I lost track of myself. In doing so I was certain I'd lost track of the mountains.

All that time, too, I was beginning to be out as queer. With queerness came the hints of community, of belonging, maybe even of love. Maybe my queerness could be the thing to knit me back together—but I didn't know how to be queer and in the outdoors at the same time.

———

I love national parks, but I have no right to feel comfortable in them. Parks were created by white men to exercise their masculinity: for hiking and mountaineering and the things that would help men feel like Real Men. Teddy Roosevelt spoke with admiration of "the hard contests where men must win at hazard of their lives and at the risk of all they hold dear." National parks were the ideal playgrounds for his "strenuous life"— and they were not considered a place for women, who were expected to remain tied to the domestic sphere. Among many outdoorsmen, this is a sentiment that remains.

Even when I've known with confidence that my wilderness skills are strong, I've faced doubts from men. The advisor for my college outdoor club assumed I didn't know how to drive in snow and that I couldn't lift a heavy backpack. Many of the boys I climbed and hiked with as a teenager were more likely to comment on how my ass looked in a harness than on my climbing abilities, and I was expected to take that as a compliment.

Though outdoor spaces are now where I feel happiest, as a small woman, I am an anomaly in them; my queerness makes me even more out of place. I've passed men on the trail that put me on immediate alert because of some small action: a stick swung at a tree branch in anger, a voice raised too high and tinged with threat. I carry bear spray not so much for bears but for men.

And yet, I also have a privilege many other queer women don't: years of experience in nature spaces, and the whiteness that keeps me from raising too many eyebrows on the trail. I don't carry the cultural memory of ancestors chased off into the mountains or lynched in trees that many Indigenous and Black people and other people of color do. The outdoors have been my playground since I was a kid. I know how to adopt the bravado that makes men accept my skills; I wear the right clothes, carry the right gear. I'm able, more or less, to fit in.

————

Throughout college, I had a print of one of Ansel Adams's photographs on my dorm room wall. It was an image of Half Dome lovingly rendered in silver and black tones. It was expansive; it was awe-inspiring. It was the pinnacle of outdoor glory.

I bought the print at a college poster sale despite having never been to Yosemite, and while I loved the photograph, I didn't think too much about it. I certainly never stopped to consider that it was missing something major: people. Since time immemorial, the Yosemite Valley—Ahwahnee, or the place of a gaping mouth—has been home to the Ahwahneechee people, who have relied on the landscape for sustenance and their way of life. But in 1851 and 1852, hundreds of Ahwahneechee people were pursued and slaughtered, their food stores burned to starve any survivors, by white men looking to clear the way for gold miners. The period is now referred to as the Mariposa War, as if it had truly been a two-sided battle rather than a massacre. By 1910, 90 percent of the Ahwahneechee inhabitants of Yosemite were dead, the rest left to piece together a life on the fringes of the invaders' encampments.

John Muir's first visit to the Yosemite Valley was years after the war ended. Camping on the shore of Tenaya Lake, he noted happily in his journal that "No foot seems to have neared it." He built a cabin along Yosemite Creek, celebrating the quiet and sense of peace he found along the water's edge, and lived there for two years. His familiarity with and love for the land—not its people—would be instrumental in establishing the valley as a national park.

So many of our national parks have the same history: the Klamath people were forced to cede lands in and around what is now Crater Lake National Park; the Blackfeet and Flathead were pressured to give up the mountains of Glacier National Park; the Shoshone, Bannock, and others were pushed out of what is now Yellowstone. Most left under treaties that stated they would still have traditional hunting rights in the new parks—treaties that before long would be broken and ignored by the United States. And it wasn't just the land that was taken, but Indigenous Peoples' stories and cultures too. In Yellowstone National Park, staff erroneously claimed that Indigenous people had been afraid of the geysers and ignored the area completely. Mount Rainier and Olympic National Parks maintained that the fear of unspecified spirits had always kept Indigenous people away, while similar excuses were made in Glacier and Zion.

Since the National Park System's inception, the myth of the national parks has been consistent: these places are wild, and because they're wild, people were never there.

I didn't know all this history when I left to go backpacking in Alaska; I hardly knew these forefathers' names. But their ideas had seeped into me through the backpacking I'd done each summer, the hiking I'd done with my dad, the climbing I'd done with my brother. I had absorbed the sense that there was land, and there was us, and the only way to protect the land was for us to leave it alone, with occasional forays in to appreciate its distant beauty. The wilderness, and especially federal wilderness designations, I thought, were important to save the land from us.

———

I am not a religious person, or even someone who believes in a god—I haven't been in a long time, or perhaps ever. But the only times I've found myself contemplating the existence of some higher power is in the mountains of the American wilderness. Sitting in an alpine meadow or walking along a knife-edge ridgeline, I have been overtaken by the eerie sense that there is something bigger, something alive.

I don't believe that "something bigger" is a sentient being plotting the fate of all mankind, or a wise old man watching whether we do good. Rather, I see the land, and everything that lives in it. The mountains I've crossed, the tundra I've trodden upon, the woods I walk through with my dog—all of that is unquestionably alive and has an existence far beyond my own.

Indigenous Peoples have understood this fact for millennia. It's in their stories, in their language. Potawatomi ecologist and writer Robin Wall Kimmerer describes the Potawatomi language as one that views the natural world as inherently living, possessing agency and action. Most things we think of as nouns in English are verbs in Potawatomi. "A bay is a noun only if the water is *dead*," she explains. "When *bay* is a noun, it is defined by humans, trapped between its shores and contained by the word. But the verb *wiikwegamaa*—to *be* a bay—releases the water from bondage and lets it live. 'To be a bay' holds the wonder that, for this moment, the living water has decided to shelter itself between these shores, conversing with cedar roots and a flock of baby mergansers."

To view the natural world—mountains, bays, lakes—as verbs sets them free from their confines as static, passive things to be used by humans. It acknowledges that without people to witness and name them, these places would continue being, continue acting; the world doesn't need us around to approve or disapprove. And verbing the world acknowledges that we are just one group of actors among many: just as we may choose where to go and how to spend our time, so, too, do our mountains, our rivers, our plains.

But when many of us white people think about these Indigenous ways of viewing the world, we have a tendency to think that viewing the world as alive is the same as viewing the world as *untouched*. We assume that to return to pre-Columbian America would be to return to wilderness as we think of it today: separate and pure; unpeopled. It would be a land where animals roamed where they wanted to; where rivers followed their chosen course; where forests burned and regrew as they felt necessary—all in perfect harmony.

But the truth is that it has been tens of thousands of years, not just a few hundred, since this continent was "pristine"—that is, human-free. By the time of European contact, most of the land now known as North and South America had been managed in some way for generations upon generations. From 900 to 1450 CE, the Hohokam of the North American Southwest maintained extensive networks of irrigation canals, with more than 800 miles of trunk lines and hundreds more of local extensions. In the Northeast, Indigenous communities burned the undergrowth in forests to create meadows, which would then attract deer, elk, bears, and other game. In the Great Plains, several different nations transformed the landscape similarly, using fire to extend the grasslands for bison and other game; on the Oregon coast, the Tillamook used fire to clear the way for berries and other key food plants. The landscape was balanced—no factories spewing effluent into rivers or pit mines scarring mountaintops—but it was hardly free of human intervention. A world in balance doesn't have to mean a wilderness without people.

In most histories of the United States, we settlers tell ourselves that before Europeans came here, the landscape was a vast wilderness, untouched and unused. We assume that because *we* weren't here, no one was. That couldn't be further from the truth.

———

European explorers in the fifteenth and sixteenth centuries abided by the Doctrine of Christian Discovery, a papal bull that distinguished

Christian lands and nations from non-Christian ones. Only Christians had the privilege of political title to lands, so any foreign lands explorers found—whether people lived there or not—could be readily claimed.

The Doctrine of Discovery enabled British, French, Spanish, Dutch, and Portuguese explorers to claim the "New World" for their own, even though Indigenous Peoples had lived there, relying on and caring for the land, for thousands of years. The nascent United States grew out of that land grab. In 1823, the Supreme Court ruled in *Johnson v. M'Intosh* that even though the lands of the Eastern Seaboard were occupied by Indigenous Peoples when Europeans showed up, the Doctrine of Discovery had automatically given Europeans the right to title and use. "The tribes of Indians inhabiting this country were fierce savages whose occupation was war and whose subsistence was drawn chiefly from the forest," wrote Chief Justice John Marshall. "To leave them in possession of their country was to leave the country a wilderness."

Through *Johnson v. M'Intosh*, the courts paved the way for westward expansion, and the popular cultural imagination followed suit. In 1845, journalist John O'Sullivan described "our manifest destiny to overspread the continent allotted by Providence for the free development of our yearly multiplying millions." To call American expansion "unrightful" or "unrighteous" was "wholly untrue, and unjust to ourselves," he wrote. It wasn't just the right, but rather the *duty* of white people to expand westward across the continent.

The idea of Manifest Destiny became key to the annexation of Texas, the Louisiana Purchase, the occupation of Oregon Country, and more. Manifest Destiny said that white people could, and must, bring Christianity and capitalist democracy to the continent. It was their responsibility to move westward and to transform the land. It was all theirs for the taking.

For a few decades after US independence, laws and treaties with Indigenous nations often barred settlers from extending into Indian

Territory. It was nearly impossible to stop settlers from taking the land they felt they deserved, though, and eventually, the US government didn't try too hard to stop them. Ultimately, it officially bought into Manifest Destiny, broke treaties, and actively encouraged settlers to move west. The Homestead Act of 1862 granted settlers the ownership of parcels of 160 acres of public land in the West for a small fee and five years of working the land. The Dawes Severalty Act of 1868 disintegrated communal Indigenous lands, forcing families onto small private plots and putting whatever was left over up for sale to white people. And the 1877 Desert Land Act encouraged white farmers to claim and irrigate lands in arid and semiarid regions of the western territories. These lands were taken, either by force or duplicitous treaties, from Indigenous Peoples, and in white people's hands, they were used in often unsustainable and destructive ways.

In the writing and discourse of the time, wilderness was often likened to a woman: virgin, pristine, and full of untapped fecundity, ready to be tamed. "Invasion [of the West] was described in highly erotic terms," writes Rebecca Solnit. "The land was virgin, untouched, undiscovered, unspoiled, and its discoverer penetrated the wilderness, conquered it, set his mark upon it, claimed it, took possession of it with the planting of his flag or with the plough that broke the plains." The land was an unsullied woman waiting to be wed, to be claimed, to be turned to whatever purpose the discovering men had in mind.

The land, in essence, was stolen twice: first from its original stewards, and then again by white men who assumed it had no agency or use beyond their planned exploitations. The land had no mind of its own, no ability to act without a man's guidance—specifically, a white man's guidance. It was there to be used, inanimate; it was there to serve.

———

I went to Alaska when I was fifteen years old, the age where whatever comfort I had once had in my own skin had dissolved completely. I was growing up, becoming a young woman, but I had few models of a womanhood I wanted to embody.

I'd always been a bit of a tomboy, but now I struggled to keep up with the boys, who were constantly trying to outdo one another—to hike faster, to climb higher peaks. And I was baffled by the girls, who seemed to speak some language I had never bothered to learn. Before long, a clique had formed, and I found myself on the outside with no idea how it had happened or what exactly made me different from the others.

I remember struggling on long days of high elevation gain, trying to keep my breathing even, trying to keep pace with the others even when I stumbled and tripped on my own feet, when my calves were exhausted from hiking across soft summer tundra that absorbed our strides like sand on a beach. I remember telling myself to hold back my tears, to hide my frustrations, to never, ever admit that I was having a hard time. I remember days where each step was misery, nothing but pain and repeating to myself that I couldn't do it, that I would never be good enough.

And yet, the whole time, I wanted desperately to be there. I loved coming over a ridge or around a corner and seeing yet another towering peak, each a different shade of iron-red or shale-black. I loved the glimpse of a lone coyote I caught while sitting at a lake in silence; loved the moose I watched chewing on greenery while up to its knees in water; loved the huge herds of caribou traveling across the tundra. I loved tracking our process on topographic maps and knowing that we were miles and miles away from the suburbs I was used to.

In retrospect, I understand that I was lonely. I was coming into my queerness. Before I came to Alaska, I had been getting weird, confusing crushes on my closest female friends that made our friendships complicated. I was constantly pushing against the constraints of teenage femininity high school was imposing on me. Here in Alaska I was the weird one, the nerd. The one who hid from conversations about the pop culture I knew nothing about by learning how to recite types of rocks and cloud formations.

I didn't know how to talk to straight girls; I wasn't really sure I wanted to talk about the things they talked about. I loved the pull I had on boys, but something about them, too, made me increasingly uneasy. And though I hadn't put my finger on all that, it must've been obvious to them. Something weird about me, something off, must've stood out.

I learned on all those backpacking trips that nature and culture are separate and distinct, two binary opposites that have no moments of overlap. At home, I had greenery and trails along creeks and rivers, but I had no *wilds*; I had to go to the unpeopled mountains to experience real nature. I knew animals and plants were living, but I was taught that they had no real culture of their own; to ascribe any of that was simply anthropomorphizing—wishful thinking.

Of course, we're finding out more and more that this couldn't be further from the truth. Just as we can't draw simple boxes around what it means to be a man or a woman, nature and culture overlap and swirl together in a kaleidoscope of possibility.

Culture abounds in the "natural" world. Humpback whales share learned feeding behaviors among their kin groups. Ants teach one another how to find food, leading each other to distant morsels. Chimpanzees use tools in ways that are unique to their own troops, passing down techniques and idiosyncrasies through families. And it's not just animals: even trees communicate, using networks of fungal mycelia to share nutrients and warn one another of pests and toxins. Perhaps none of these is exactly the same as human culture, but the barrier between nature and culture is becoming more and more permeable the more we learn.

When I was coming out as queer, a debate raged around the nature–culture binary. Were gay people gay because of genetics—we're just born that way? Or is it that we learned to be gay, or perhaps just *want* to be because it's trendy? (Today, a similar conversation swirls around trans identities and gender, but then, mainstream culture hadn't caught up yet, and many gay advocacy organizations seemed content to keep it that way.) The distinction felt to many of us like a matter of life and death—our very rights hinged upon this question. Conservative Christians aligned homosexuality with culture: we were degenerates, choosing to be immoral with our sodomitical tendencies. In response, many gay people leaned into the idea that homosexuality is something written in our genetic code. We were gay because we *had* to be, not because we wanted to be, so it would be wrong to discriminate against us. Homosexuality was inscribed on us by nature, the logic went, so the world needed to treat us as natural. Numerous court cases seeking gay rights depended on homosexuality being a biological fact.

A less vocal, or rather, less mainstream, part of the LGBTQ movement wondered if the nature–culture binary was a false one: if genetics could predispose us to being attracted to the same sex, *and* our experiences could further shape those attractions. We could be influenced by our biology but also by our families, our friends, the world around us; the boundary between the two was permeable, difficult if not impossible to pin down. Maybe we could have it both ways; maybe that could be enough.

———

Most queer people grew up not knowing many other people like us, feeling out of place, aberrant, alone. With that in mind, many of us congregate in highly populated places, places where we're able to find one another. Places where we're visible. That may be changing as young queers experience unprecedented levels of acceptance among their peers and parents and can connect with one another via social media, but at least for my generation and older, the city is where it's at. Washington, DC, where I grew up, is brimming with gay men, as is Seattle. Minneapolis is full of queer women. When I first moved to Portland, I was giddy seeing all the queer people of all genders around me.

But white queer people are also among the great gentrifiers. Historically, we've made less money, had less family support, and been more willing to live in "dangerous" areas for the sake of cheap rent than our straight counterparts. That means we're quick to move into "less desirable" neighborhoods—that is, neighborhoods populated by people of color, and by poor and working-class people.

When my ex and I lived in Seattle, we lived on the edge of the city's sole historically Black neighborhood in the converted attic of a house owned by a white gay couple who had lived there for about a decade. It was one of the only places we could afford that was close to Seattle's gay neighborhood and didn't require a long commute for either of us.

We were among the only five or six white people on the block; everyone else was Black. But over the several years we lived there, our landscape changed entirely. Two houses down, our neighbors sold their house to a white couple with two young kids for half a million dollars. Then our next-door neighbor, an elderly Black woman, outlived her reverse mortgage and lost her house. It was bought, renovated, flipped, and sold to a white couple. Month by month, the houses changed over, the neighborhood growing more and more white, a clear divide showing up between the Black neighbors who all knew each other and checked in on one another and the white neighbors who only, it seemed, talked among ourselves. All the while, businesses that catered to us popped up throughout our neighborhood—coffee shops, yoga studios—and rents climbed. By the time we left, our block was almost all white, and had our landlords been charging us market rates, we never would have been able to afford our apartment.

Gentrification is often seen, particularly by white and wealthy people, as a good thing—a revitalization. But what's often overlooked is who it displaces, and why those neighborhoods were "undesirable" or run down in the first place. The whole process, Eula Biss points out, "betrays a disturbing willingness to repeat the worst mistake of the pioneers of the American West—the mistake of considering an inhabited place uninhabited." We were just looking for a cheap place to live, one that we could afford on my grad school stipend and my ex's nonprofit salary. We weren't trying to kick anyone out. But our very presence contributed to a shift in the balance, and it wasn't as if we were doing much to bridge a divide; I don't think we talked to the Black families any more than the other white people around us did. Without meaning to, we were colonizing.

We were, essentially, repeating the process that created the national parks, erasing the people who had lived there so we could have our space the way we liked it. We didn't ask for a coffee shop or a yoga studio, but when they opened, we were happy to visit.

———

The mountains in the Pacific Northwest and throughout the United States bear the names of white people. When I lived in Seattle, there was Mount Rainier, named for British Royal Navy rear admiral Peter Rainier. Now I live in the shadow of Mount Hood, named for another British naval officer, and Adams and Jefferson, both named for presidents. And Mount St. Helens, named for yet another British man.

Perhaps we can cut early settlers some slack. They were trying to make these foreign places in a foreign land more legible to themselves, to create cues that reminded them of home. But renaming also suggests an obliviousness, and a lack of willingness to take as legitimate the history and culture before them. "We all might do well to remember," writes Lauret Savoy, "that names are one measure of how one chooses to inhabit the world."

All these mountains had names before we showed up, names given to them by the Nisqually and Salish, the Chinook and Cowlitz, the Molalla and Kalapuya. To the Multnomah and the Klickitat, there are a number of versions of the story behind the mountains. In one, Mount St. Helens was Loowit, a beautiful young woman. The two sons of the Great Spirit, Wy'east and Klickitat, fought viciously over her, raining down rocks and trees and destroying the great bridge over the Columbia River, Tahmahnawis, in the process. Furious at their violence, the Great Spirit turned Wy'east and Klickitat into mountains—what we now call Hood and Adams—and Loowit too.

To erase those place names is to erase their stories, and in many cases in a way that is irreparable.

Many of the Indigenous Peoples of the Northwest—and the rest of North America—were killed within a generation of encountering white people. They were slaughtered in raids and in warfare, and wiped out by diseases brought by settlers. Those who survived were forced through a system designed to "assimilate" them into white culture—that is, to halt the transmission of their cultures from one generation to the next. Communities were forced off their homelands and onto reservations that had scant resources and often required methods of hunting, ranching, or cultivation that were unfamiliar to them. Children were taken from their parents and sent to boarding schools where they were abused, forced to speak English and forget their native languages, forbidden to practice their own religions and made to convert to Christianity. The schools were so dangerous, so rife with disease, starvation, and abuse, that many children died there and were buried in unmarked graves. Entire generations had their lives, cultures, and memories stolen from them, and while many communities are attempting to rebuild, some of what they once had is lost forever.

When we call their mountains by the names of distant white people, we perpetuate this erasure.

———

As a child growing up on the East Coast and attending public schools, I was taught that most Indigenous people were gone. That's how it was put, as if they'd just decided to pack their bags like the elves of Middle Earth and sail on to another land. We learned about the Wampanoag and the Pilgrims and a Disneyfied version of the first Thanksgiving, and then virtually never heard about Indigenous Peoples again.

What was unsaid, except perhaps for a vague and cursory gesture toward the Trail of Tears, was that Indigenous people here were slaughtered, hunted, packed off like livestock to places completely dissociated from their cultures, histories, and foodways. We learned, instead, that they had just died out, innocently and incidentally—vanished.

"*Vanish* is a deceptive word. It slips easily off the tongue, the soft *sh* a finger to the lips quieting a history far from simple, neat, or finished," writes Savoy. "Fragmented, dispossessed of land, dislocated, perhaps ravaged by disease and violence, tribal people endured. Members reorganized or joined other groups. They migrated or they stayed in smaller communities. They continued to speak." But where I grew up, we didn't hear their voices.

It wasn't until I drove across the country, moving to Seattle in my early twenties, that I began to realize just how wrong my sense of history was. I drove past and through reservation after reservation, pockets of sovereignty I had somehow thought were gone forever. I felt like a fool for thinking these cultures had vanished.

It was disorienting, a little embarrassing, and even more embarrassing was that I had absorbed a sense that these places were dangerous. My girlfriend at the time and I drove quickly through reservations, hesitant to pause and talk to anyone. We were unwilling to camp on the federal lands neighboring one reservation, even when the evening light was fading and we were exhausted and hours from our next planned stop. As far as we knew, these lands were a continuation of the Wild West's Indian Territory, full of peril and risk.

I knew at the time that my hesitation was probably racist and without justification, but I couldn't overcome it. We kept driving.

The vanishing and savagery of Indigenous Peoples is a story we've told ourselves for ages, one that forms the bedrock of this nation. It's how we've justified our endless land grabs and the genocide we've committed in the name of expansion. I've benefited from this narrative, as have so many of us, and I am trying, now, to be diligent. I'm trying, now, to tell a different story.

———

One morning when I was on that trip backpacking in Alaska, I woke early, my turn to make breakfast for my tentmates. It was quiet, the only sound the small creek burbling beyond our campsite, and the sun was out. It had rained for most of our trip, and it was one of our first clear mornings, with blue sky stretching all above me. I could see the expanse of tundra below; in the distance, a herd of caribou crossed a valley.

We were camped midway up a slope on a small, flat area. My eyes were bleary with sleep when I looked out at the distant mountain ridges, and my first thought was, Huh, that's an interesting cloud.

It wasn't a cloud. It was Denali, a stark-white mountain towering above the other peaks, easily two or three mountains taller than the rest. It was like a giant, sleeping polar bear, nestled on the roof of Alaska.

Denali isn't just *a* mountain; in the landscape of south-central and central Alaska, it is *the* mountain. In Denaakk'e, the mountain is Deenaalee, "the high one"; in Dena'ina—the language of the people in whose territory the mountain stands—it is Dghelay Ka'a, "big mountain."

When I was there in 2004, glimpsing it in my early morning confusion, officially it was still Mount McKinley. That name had been on the books for more than a century: in 1896, a gold prospector named William Dickey had glimpsed this enormous mountain and felt it was his right to give it a name. He named the mountain after the current presidential nominee and future president.

The name stuck until 1975, when the state of Alaska changed the mountain's official name to Denali and requested that the federal government do the same. But that kicked off a forty-year debate over who had the right to name it: the state of Alaska, in which the mountain stood, or Ohio, the home of President McKinley. Ohio Congressman Ralph Regula introduced legislation every two years demanding that the mountain's name not be changed, effectively tying up any efforts to honor Alaska's wishes in bureaucratic motions. That worked until 2015, when finally, amidst protests from Ohio, Interior Secretary Sally Jewell stepped in and changed the name.

The insistence on the name Mount McKinley had been an attempt by Ohio to keep the focus on their state's history and on the legacy of President McKinley. But it also said something else: that the supporters of the name Mount McKinley didn't think the first people of Alaska deserved the right to name their homeland. It said, essentially, that people three thousand miles away knew the mountain better than the people who had lived around it since time immemorial.

———

Early morning on the last Christmas of the Obama Administration, Leslie and I went snowshoeing in Mount Hood National Forest. By all rights it should have been a miserable hike. It was cloudy and gray, and I was in the waning days of a bad winter cold that had knocked me out of commission for more than a week while the rattle of a cough in Leslie's chest was evidence that they had caught it from me. But the numbing winter air and a steady stream of cough drops soothed our throats, and the sight of Goose, our dog, twelve years old but still running wind sprints up and down the trail and plowing her face into snowbanks with glee, made us continue on.

We'd gone up there to seek refuge. Just weeks before, Trump had been elected, and it felt as if our world had shattered. We were both living in a constant fog of depression and fear, and we didn't even know what was to come. We didn't know that in the first six months of his presidency, Trump would hasten permits for oil pipelines, appoint an oil baron as his secretary of state, reassign climate change staff at the Environmental Protection Agency and scrub federal climate change websites, expand offshore drilling, pull out of the Paris Climate Agreement, order a review of all national monuments, and more. We didn't know he would roll back LGBTQ workplace protections, stop schools from protecting trans students' bathroom use, and nominate an anti-gay justice to the Supreme Court. All we knew was that whatever the new administration was going to do wouldn't be good, for us or any of the places or people we cared about.

We were almost at the viewpoint when the sun began to come out. Through the trees, the knife-edge of the mountain's peak appeared, jagged and crisp above gray tufts of clouds. We trod on past snowy trees, our jubilant cries interrupting the rhythmic thunking of snowshoe against boot, the soft plunge of pole into snowbank. Small birds squawked and creaked at us as we passed the trees they'd gathered on. By the time we reached our stopping point for lunch, the clouds had blown away, leaving the whole peak visible, splendid in its snowy finery.

It felt like a cliché; it felt like it had happened just for us, a reward for getting out of bed early on that day off. I felt an echo of the elation I had felt the first time I'd seen Denali, but now, so much had changed.

Here we were on stolen land, two queers in love seeking sanctuary on Christmas from a world that threatened to continue its theft, on a mountain that felt like home. I don't know how to hold all those things in my hand at once, but I know that I must.

To heal this world, I have to know that Mount Hood is my home, but it is also Wy'east, a mountain stolen from people who have suffered and endured, who still have legitimate claims to these places today. I have to know that as Eve Tuck and K. Wayne Yang write, "Decolonization is not a metonym for social justice." True unsettling doesn't just mean hoping things get better or educating ourselves a little bit here and there. It means *land back*. It means change, and not just incrementally.

Here, we are but visitors.

4

SURVIVAL

On a windy, chilly February morning in 2002, a lone coyote trotted his way across the tarmac at Portland International Airport. He skirted the concrete, then scurried into the grass, looking for mice and other snacks, any meals that might be hidden away among the greenery.

He was probably not the first coyote to visit the airport: its open expanses of grass no doubt harbor all sorts of rodents, a boon to hungry canids. But this coyote had a different destination in mind. He jogged past parked airplanes and around vehicles, dodging airport security and oblivious travelers as he crossed the entire airport campus.

Finally, tired of the airport or perhaps just looking for a quiet place to rest, he took refuge on a light rail train headed downtown. There, he nestled into a seat, nose tucked neatly under his tail, until forced to disembark before the train left the station. He was ushered away to somewhat wilder areas near the Columbia Slough, off to find a new adventure.

In rural Indiana—incidentally, just a five-minute drive from where Leslie's mom lives—you'll find the only coyote-specific animal rescue in the world: the Indiana Coyote Rescue Center. It's an odd place for it, dropped in the midst of endless cornfields an hour or so outside of Indianapolis, a series of enclosures behind a farmhouse in a region not known for its fondness for coyotes.

Coyotes range throughout Indiana, from farmland into the cities. The rescue center isn't trying to sustain an ailing population—rather, it's there because people tend to react in one of two ways to coyotes. They try to kill them, or they try to keep them as pets.

When a coyote is injured or a little too used to people for anyone's safety, this tiny sanctuary takes them in. A coyote pup adopted as a pet will still grow up to be a semiwild animal with instincts to hunt and roam. Kept as an adult, it may hurt the people around it. Setting a former pet free isn't much better, though. A coyote that's used to humans may approach strangers for food, sometimes aggressively; it will also lack its usual aversion to dogs and get into fights. The rescue center keeps these habituated coyotes safe and fed and gives them companionship with other coyotes like them.

Leslie's mom Deb lives on a cattle ranch in rural Indiana. We were visiting for Christmas when the conversation turned, as it often seems to in farmland, to coyotes and whether they are a scourge or simply benign. Deb mentioned offhand that there happened to be a coyote rescue center just down the road. It turned out to be closed to the public, but when I called and told them I was interested in coyotes, the director, Jami Hammer, invited me for a visit.

Leslie and I drove past winter-barren cornfields to the rescue center the next day, a sunny day so cold it hurt to be outside. With wind chill, the temperature was in the single digits, and frosty air numbed my face in seconds. I'd packed only thin boots for our visit, little help against the snow and ice. But coyotes have no problem with cold. With thick coats, they curl up and put their backs to the wind and hardly feel a thing. When we got there, they were romping around in the snow like it was a balmy summer day.

I'd always thought coyotes were at least German shepherd size or so, but they're surprisingly small. By weight, they're only about half the size of a timber wolf—smaller than a golden retriever, and about the same size, though slimmer and more lithe, as my forty- and fifty-pound herding dogs.

Beyond their size, what's most striking about a coyote is its gaze and the sounds it makes. On the street or in the mountains, a coyote will lock its uncanny eyes with yours, at once wary and scrutinizing. And where wolves howl in deep, haunting harmonics, coyotes create an odd, unsettling collection of yips, screams, and barks. While wolves on a cold winter night might sound like ghosts whooshing through the trees, coyotes are the banshees of the forest.

Coyotes are often curious, and many of those at the rescue center were no exception, watching us as we passed. We were warned to keep our hands a good distance from the fences lest an inquisitive coyote snatch our phones away from us or sink sharp teeth into our fingers. A few trotted toward us, sniffing and inspecting us through the fence.

One young coyote, Neegan, gallivanted through the snow in his enclosure with a stuffed animal. A bonded pair, Artemis and Orion, yowled at us, then negotiated mealtime the way, apparently, they always did: Artemis ate her fill, snarling at Orion to keep away until she was finished with the meat. Other coyotes were nowhere to be seen, shying away behind bushes or inside doghouses, waiting for us to leave.

One coyote, Ares, let me feed him canned peas from a spoon. Jami and I entered his enclosure, and I crouched a few feet from the coyote. His coat was thick and full, a deep, lush mix of black and grey and brown. His eyes were wary and curious, holding my gaze with hazel intensity.

As I crouched there with him, the chill of the winter day slipped away. I was there with an animal I'd read so much about, so close that I could have reached out and buried my fingers in his thick fur. Instead I scooped out a spoonful of peas and held them toward him, keeping as still as I could. I held my breath as he inched closer and closer, slowly beginning to trust me.

———

When I lived in Seattle, my girlfriend at the time and I hiked most weekends in the Cascades an hour or so outside the city. On one midsummer day, we chose a route that meandered through a valley before arcing up to a saddle between two peaks. In the valley, the wildflowers and berries stretched above our heads so thickly that we couldn't see past the next bend in the trail, bees and other bugs flitting around us in search of sweet summer nectar. I wanted to relax, but the fresh growth made it prime bear habitat, all too likely that we'd come around a corner and see a black bear foraging with her cubs and ready to see us as a threat. We held our bear spray out, talking excessively loudly so that any blissed-out bears might hear us coming and head the other way, and listening carefully for any tell-tale snaps of twigs.

After an hour of tense, quiet hiking, the trail steepened, and the meadow gave way to shorter grasses and flowers. Relieved, we stowed our bear spray. We followed switchbacks up to the saddle, ready for lunch and a rest.

The air was warm and dry, and we could see all the way to an alpine lake below. I was halfway through taking my backpack off when I looked at a patch of late-summer snow still nestled against the rocks. Smack in the middle of it was the biggest paw print I had ever seen. It was three or four inches across, far too big to have been left by a dog.

I let my pack drop with a thud. A chill ran from the crown of my head down my spine as I imagined the creature large enough to leave that print.

It was at least a few hours old, melted and blurred at the edges. I could imagine the cougar who had left it pausing here as we did, surveying the rocks and the lake below, before heading down the slope to the water to drink or maybe find a meal.

At least, we hoped it had headed that way. There was always a chance it had stuck around, had climbed further up the mountain and was watching us now. My girlfriend and I were tired and ready for a break, but we hardly had to discuss—we'd be eating our lunch elsewhere.

I assume that afternoon was the closest I've ever been to a large predator in the wild, but the odds are good that I've been closer. The woods here in the Northwest are thick with moss and trees, and cougars are stealthy, coyotes watchful. There's a good chance that the spots my dogs think smell so interesting are places marked by wild animals. There's a good chance that the feeling I get sometimes that I'm being watched isn't just a feeling.

———

Since seeing that lone cougar footprint, I've started paying more attention to the prints I see when I'm in nature. When hiking, I remind myself to look not just at the views but at the ground too, for hints of what's come before me. In a palm oasis just outside of Joshua Tree National Park, I crouched on a boardwalk to examine raccoon prints in the mud, a perfect trail of tiny almost-human hands arcing toward the clear trickle of water running among the palms. In the desert nearby, I smiled when I saw an unmistakable coyote print, the sharp claw marks and narrow space between the pads—and the park's ban on pets—confirming it was no domesticated dog.

Near where I live in Portland, there's a wetland preserve, the perfect place to see bird tracks and deer prints. Once I saw a track that befuddled me for hours—almost half-human, half-canine—until I got home and the internet told me it belonged to a river otter. At the beach, I search for signs of shorebirds hopping along while they dig for a meal; on the trail in the mountains, I'm always on the lookout for odd prints or scat. I've seen elk tracks on Mount Hood, ground squirrel and coyote scat in eastern Oregon, crow and gull prints on coastal beaches. I hope, one day, to see the pebbly footprints left behind by porcupines in snow.

Like many humans I am loud and clumsy, and my presence tends to shoo animals away. Prints left in mud or sand help me glimpse who lives around me when I'm in the wild. Footprints tell me whether I should keep a watchful eye for predators or if my companions are more docile beavers or opossums.

Humans leave prints too—on muddy trails I follow the tread of high-tech hiking boots, while through barefoot beach prints I can see who has flat feet or high arches, who was running and who was just wandering along the wrack line. But the footprint that gets the most notice these days is our carbon footprint, the heavy steel-toed boot we're using to tread upon our planet. The United States emits more than fifteen metric tons of carbon dioxide per person every year. That's twice as much as China, and almost ten times more than India. That carbon footprint is what's causing this climate crisis, warming the planet, disrupting our weather patterns, and provoking mass extinctions.

I like to think my footprint isn't as heavy as others'. I take public transportation whenever I can; I've been vegetarian since I was a teenager; we pay extra to our utility company to subsidize renewable energy. But I drive an hour or more to the mountains most weekends to hike; I fly to far-off destinations for work and for fun. If someone were tracking my movements in some metaphorical woods, they wouldn't think I walk on two feet—they'd think I was riding a bulldozer.

———

As we humans sweep across the land and make our mark upon the landscape, the line between non-urban and urban grows ever thinner. We've taken over enough space that many animals are moving into our cities out of necessity or opportunity. You don't have to go to the woods to know that some creature has an eye on you, or at least is interested in rummaging through your trash.

Today, coyotes have made their homes throughout North American urban areas. In New York City, they live in Central Park; in DC, they den and hunt in Rock Creek Park. They pop up all over the place in Chicago—up to two thousand coyotes are estimated to live in and around the city.

In Denver and Portland and a few other cities, coyotes are getting too used to people. Many of us who aren't accustomed to seeing wild predators in our neighborhoods find coyotes cute and enthralling in their dog-like ways. Our habit of watching them in quiet awe rather than chasing them off—and our perpetual bounty of discarded meals—has made some urban coyotes bold, increasing the likelihood that they'll approach humans for food or out of curiosity. And as the Indiana Coyote Rescue Center can attest to, a coyote that's used to humans is a coyote that may find itself in trouble.

Here in Portland, coyotes are thriving: more than eighteen hundred sightings were reported in 2018 alone. Every few weeks, a panicked post goes up on my neighborhood's discussion board warning people to keep their pets inside because a coyote's been spotted. About once a year, one appears like a ghost out of the dim dawn light when I'm on my morning run, then disappears, loping off among the houses.

———

Among the Colville Tribe of Eastern Washington, Coyote is known for causing the Missoula floods and creating the Columbia River,

N'chi-Wana. At first, the story goes, there was a great, vast lake up in the mountains. Coyote was unsatisfied that the water was locked up and static—he wanted there to be a river so the water could be shared with the world. So he dug and sang, dug and sang, until finally the ice and rocks gave way and a great river rushed forth. That river gave life and sustenance to all the people who would come to dwell beside it.

Stories of Coyote abound among Indigenous Peoples from the Southwest to the Great Plains, wherever coyotes lived alongside people. In these stories Coyote is both human and animal, trickster and helper, wearer of many faces and genders and multiplicities. Many of the stories are tied to certain places or times of year or can only be told by certain storytellers. The story of Coyote creating N'chi-Wana was shared with me and other artists in 2018 by Ed Edmo, a Shoshone-Bannock poet and storyteller who grew up along the river.

To the white people who made their way to this continent, coyotes and other predators were one thing, and one thing only: a threat. Their tendency to hang around camps, and their uncanny yips and howls, unnerved settlers making their way across the Great Plains. When settlers established ranches and began to run cattle and sheep across the continent, coyotes became the figurehead of not just all that was strange and alien about this new home but also of all that was hostile. Settlers believed coyotes and their wolf cousins slaughtered livestock, and the only way to protect the livestock was to do away with the predators. "From the 1880s until the 1930s, the received wisdom in America, very rarely questioned, was that the only good American [coyote] was a dead one," writes historian Dan Flores. "The real question was how to kill as many coyotes as possible in the very shortest period."

Ranchers and the United States government demonstrated truly impressive creativity when approaching the problem of how to get rid of the coyote. There were traditional hunting techniques like trapping and shooting, of course—but that was only the beginning. In the 1830s, coyote killers began to rely largely on strychnine, a potent poison that targets the central nervous system.

Strychnine acts quickly, causing painful, horrifying cramping and convulsing that can last for an hour or more. In just a few hours, or after up to a day of pain, the convulsions and seizures kill the animal that has consumed the poison. It dies because it can't breathe or because it just doesn't have the energy to live anymore.

Those out to destroy coyote populations laced carcasses of cattle, deer, and other animals with strychnine and waited patiently for coyotes to find them. Bounties set by state legislatures made killing coyotes not just tempting for ranchers and others but lucrative—between 1883 and 1928, for example, Montana paid bounties on 886,367 coyotes at about a dollar per pelt, or about twenty dollars each in today's equivalent.

Coyotes did eventually wise up, avoiding carcasses that smelled of strychnine after watching their companions die near other baited carcasses. But then came more poisons, like Compound 1080—sodium fluoroacetate—and thallium sulfate. These alternative poisons took longer to act, making it hard for coyotes to understand the link between baited carcasses and their dead companions. But the new poisons were equally lethal and equally grotesque.

This went on for decades: between 1945 and 1971, up until strychnine, thallium sulfide, and Compound 1080 were finally outlawed due to their broader environmental impacts, the Department of Agriculture collected the carcasses of 3.6 million coyotes. To put that in perspective, about 3.6 million people live in the state of Connecticut today. Plus, an additional three million coyotes may have been poisoned but never collected. And while poisoning carcasses may have been an effective way to kill coyotes, it also had ramifications throughout entire ecosystems. It's not just coyotes that ate the carcasses. Wolves, bears, eagles, condors, and others all were likely to feed on them and suffer the painful, and lethal, consequences.

———

Among the carcasses that were poisoned in an attempt to eradicate coyotes were thousands of American bison. Before white people showed up, some thirty million—if not more—bison ranged across the continent, migrating between the Rocky Mountains and the Appalachians. To the Oceti Sakowin, the nation including Lakota-, Dakota-, and Nakota-speaking peoples, the bison are Pte Oyate, the buffalo nation. The individuals of Pte Oyate have their own agency, their own role in the ecosystem; to the Oceti Sakowin, the bison are kin.

Bison were key to the cohesion and strength of the Oceti Sakowin, and white settlers knew it. After attempts to dispossess Indigenous Peoples of their lands via war and treaty failed, settlers turned to the bison. Beginning in 1865, the US military sponsored campaigns to kill them in droves. Unlike the Oceti Sakowin, who used the animal for both meat and hides, the white people took the hides but left everything else to rot. What's more, they laced the carcasses with strychnine.

Because why not kill two birds with one stone? Here was a chance for settlers to rob Indigenous people of their food source and rid themselves of the hated coyote all at once. And if a starving Indigenous person should attempt to eat the rotting, poisoned meat, well, all the better.

Barry Lopez describes the eradication campaigns against wolves and coyotes as "the violent expression of a terrible assumption: that men have the right to kill other creatures not for what they do but for what we fear they may do." To the newcomers to this continent, there was a chance that coyotes and Indigenous people might get in the way of exploiting the land, and that simply wouldn't do. Better to eradicate them than to risk the possibility that they might cause trouble.

But I would take Lopez's claim one step further: we feel we have the right to kill other creatures and people because we know our own capacity to harm, and we don't want to feel that trauma at someone else's hand. We experience a brief, flickering current of empathy, then turn away from it as fast as we can.

———

Trauma isn't just a one-time thing, a singular moment or a recurrence that inscribes itself on a single body or mind. It lasts through generations. It can encode itself in our DNA, in the proteins that influence what we look like and how we act, what our children look like and how they act. Our ancestors' trauma can make us more susceptible to illness, both mental and physical. And trauma isn't just passed down through our bodies but through our cultures too; through our stories and our habits.

In cities, trauma crops up in the white-painted ghost bikes, tributes left where cyclists were hit and killed by cars, in corner memorials where people were shot and killed. We carry trauma in street names, commemorating those who gave their lives for justice, and in neighborhood design, when city planners determine which neighborhoods get built up in the aftermath of a disaster. And we repeat trauma, inflicting it on generations to come through geography: our cities are littered with statues of Confederate leaders, streets and neighborhoods named for presidents and generals who slaughtered hundreds or thousands of Indigenous people and ordered the internment of others.

Trauma is a thing we feel, yes, but it also a thing we carry—one that's hard to move away from.

———

I've suffered from depression and anxiety almost my entire life. I've carried sadness so molasses-thick that I felt like I walked through the world unable to make eye contact with anyone or anything; anxiety so hefty I hyperventilated at the idea of flying, or spending time in a group of people, or just leaving the house. I've flaked on hangouts and cancelled vacations because my brain felt locked up in stone, incapable of doing more than the very basic.

As a queer person, I'm hardly alone in this. Lesbian, gay, and bisexual people are twice as likely as our straight counterparts to experience mental illness, including but not limited to depression and anxiety, and trans people are almost *four times* more likely than cis people. Queer folks are at a higher risk for suicidal thoughts and attempts, for homelessness, for substance abuse.

It's especially bad for queers in their teenage years, which is when I first started experiencing my symptoms. Lesbian, gay, and bisexual youth are more than twice as likely to feel sad or hopeless than their straight peers, and almost five times more likely to attempt suicide or engage in self-harm. And as with their adult counterparts, trans youth have it even worse.

Is it so surprising that we have trouble maintaining our mental health? Many of us spent our formative years hiding, unwilling or unable to share essential parts of ourselves with our friends and family for fear of ridicule or worse. We've been harassed in the streets; many of us have been bullied by our peers. Bill after bill has been introduced in state legislatures barring trans people from bathrooms, from homeless shelters, from sports, from medical care, and more. And while numerous campaigns over the past decade or so have aimed to show queer teens that their lives will get better, it remains rare to see queer characters on TV whose storylines don't end in death, rare to see mainstream public figures embodying queerness. Marvel, for one, made it through the entire eleven years of the Infinity Saga without having a single openly queer main character. Disney continually relegates gay characters to one-line, blink-and-you'll-miss-them moments that can easily be cut for conservative audiences, and they won't go near the idea of a trans character. When we *do* get openly queer characters, most of the time they end up dead, abused, raped, or all three. *Buffy the Vampire Slayer, True Blood, Battlestar Galactica, Orange is the New Black, House of Cards, The 100, Game of Thrones, Doctor Who*—many of which were at first lauded for their inclusion of queer characters—have all killed off queer characters, often in brutal or perfunctory ways. If those are the models we have, it's no wonder that many of us live at least part of our lives under clouds of impossibility, erasure, even terror.

We also belong to a community that has, frankly, been through quite a lot. Throughout most of US history, queer identities and acts were illegal. In the early- to mid-twentieth century, when homosexual and trans identities were solidifying in the public eye, queer people dealt with condemnation on all sides. Queer people could go to jail—where they were then likely to be verbally and physically abused—for wearing more than two or three articles of clothing of the opposite gender. Gay and lesbian clubs were regularly raided by police, and queer people beaten and arrested. Being openly queer meant the likely loss of jobs, friends, and family. For the most part, our queerness had to exist in the shadows and in our hidden community spaces.

Then in the 1980s and '90s, we lost a generation to the AIDS crisis. AIDS deaths weren't reliably counted at the beginning of the epidemic due to a lack of attention from the medical and public health community, and many of the patients who *were* counted weren't out as gay. So we can't know just how much of the gay community was lost to HIV. But we do know that in New York City in 1989, AIDS was the leading cause of death among men aged 25 to 44. At its debut in 1987, the AIDS Memorial Quilt, which contains a panel for each person lost to AIDS, contained nineteen hundred panels and weighed three and a half tons.

AIDS was a trauma for an entire community and continues to be so. Many of those who would be our queer elders today died in their twenties and thirties, and those who survived watched their loved ones die horrible deaths. And the marginalization continues, with gay bashing and anti-trans legislation and closeting and conversion therapy and everything else. We survive and endure, but with the weight of decades of trauma upon our backs.

———

Despite our frequent and painful attempts to eradicate them, coyotes' spread across North America follows our own like a shadow following a body as it walks along the plains. Coyotes first evolved in the American Southwest. They followed Indigenous settlements, moving eastward, westward, and southward with people who understood them as part of the living landscape. When white people arrived, coyotes spread further than ever before, following loggers into the Pacific Northwest and fur trappers into subarctic Canada and Alaska. Eventually, they spread across the south and east too—they now range across New England and Quebec, and have been seen as far south as Panama. In the United States, the only state they've yet to appear in is Hawai'i.

Wherever settlers and coyotes have coexisted, so too has strife. But no matter what we do, they adapt.

Kill a bunch of coyotes in a given area and somehow the remaining coyotes *know*. They sense the prey going uneaten by their eradicated brethren, or maybe they just see the absence of their former companions. Instead of their usual two or three pups, they'll have seven, eight, nine—as many as nineteen—in a single litter. Their families spread, create new families, find new homes. Kill a few coyotes, and you'll have a new generation of young ones ready to take their place.

Like wolves, coyotes can and do live in packs when possible, since larger groups can bring down larger prey. But unlike wolves, if you destroy the coyote pack structure by killing a few individuals, those left behind do just as well. They split off into loners or pairs and focus on smaller prey: mice, rabbits, insects, berries.

So where destroying packs and killing pups made it possible for hunters and ranchers to almost completely eradicate wolves from the Lower 48, for coyotes, it was just another change.

————

In 2018, I went to the Bay Area for the LGBTQ Outdoor Summit, a gathering of queers to talk about inclusion in the outdoor industry. The outdoor industry hasn't historically put in the effort to represent or consider the needs of queer folks—or any non-white-cis-male community for that matter—and most of my encounters with the outdoors growing up had been with straight people. This was the first time I'd been in a space where I didn't have to choose between being queer and loving being in the outdoors. I was with people who, like me, were both. I felt held, accepted. For once, I was with people who spoke the same language I did, who understood the feeling of rupture of having a foot in both of these worlds.

We were staying in the Marin Headlands in Golden Gate National Recreation Area, perched above the shoreline just across the bridge from San Francisco. Each morning, I left my dorm room and breathed the salty ocean air deep into my lungs. Mist drifted off the water, shrouding seabirds and shorebirds before revealing them once again as the day wore on.

The first afternoon, I wandered down the hill toward the ocean and stopped just before crossing the street that cut between our dorms and the beach. There, basking in the sunlight by a picnic table, was a single coyote.

I stilled, quieting my breath so as not to disturb her. She was the biggest, fattest coyote I had ever seen, better fed even than the coyotes I'd seen in the rescue center in Indiana. Between the wild prey within the national park and, no doubt, plenty of trash and food scraps left behind by visitors, she probably had no issue finding food. And she was beautiful, her russet coat shining in the sunlight. Happiness surged through me—a wild coyote, here with me in this queer outdoor utopia.

She locked eyes with me, and it was abundantly clear that she couldn't have cared less about my presence. I knew I should yell and make a scene to scare her off, but I couldn't bring myself to. This was her home even more than it was mine.

Over the next few days, I saw several more coyotes wandering the wetlands adjacent to the coast, and each time, I felt the same burst of joy and awe. The last night I was there, we had a bonfire just outside our dorms, and a lone coyote went trotting past us, unbothered. A little while later, I went to bed while others stayed up. About to turn out my light, I heard whoops of wild yelling and elation. It took me a long moment to realize that the sounds I heard weren't my newfound queer community, but another one altogether, a band of coyotes celebrating the night.

———

Golden Gate National Recreation Area is on the ancestral lands of the Ohlone and Coast Miwok peoples. In the late eighteenth century, Spanish colonists arrived in what is now known as California, driving livestock across the land and overgrazing the lush grasslands Indigenous people depended on for food. The subsequent ecological and cultural disaster pushed those Indigenous people who survived to the missions, where they were essentially enslaved. Today, their descendants survive, working to keep their cultures alive in the face of erasure.

We had begun the LGBTQ Outdoor Summit by acknowledging that we were on these ancestral lands, and welcomed a talk from Tongva-Ajachmem artist and writer L. Frank. But neither action changed the fact that this space in which so many of us felt held and seen was possible because we were on stolen land.

I have tried to think about what would be enough. It is not enough, I know, to just try to change the conversation, to move away from the colonial, conquest-laden language of *peak bagging* and *conquering summits*, to recognize that without the death and disenfranchisement of hundreds of thousands of Indigenous people, there would be no space for outdoor recreation, queer or otherwise. I have tried to consider what it says about me—how complacent I am—that I can go to a site of literal genocide and feel that I have found my family. I am trying to live in this contradiction, to keep doing better, but I am still trying to figure out what *better* is.

———

Coyotes have no comprehensive legal protection. There are laws about *how* you can kill a coyote—you can't poison them anymore, and there are state-level restrictions on certain types of traps—but there are no laws about *whether* you can kill them. Even the laws about how you kill them have more to do with the methods' environmental repercussions than about the horror they inflict on the coyotes themselves.

In recent years, we've killed about five hundred thousand coyotes a year. Some are shot by ranchers, others "controlled" by the Department of Agriculture's Wildlife Services division. Still others are chased down for sport with hunters and packs of dogs. Some states still offer, or at least allow, bounties for killed coyotes.

Animals have more complex emotional lives than we tend to give them credit for. Coyotes have proven to be incredibly resilient, bouncing back no matter what we throw at them. But we've killed their companions and family members in gruesome ways, destroyed kin networks, and run them off their habitats. How can we assume that that hasn't had any effect at all?

When possible, coyotes form tight-knit families; they watch out for one another. Older juvenile coyotes tend to serve as extra parents to their younger siblings. Coyote pups and adults play together, which in addition to helping cement hunting skills and familial bonds, also just gives them the opportunity to experience something akin to our own joy and fun.

Coyote grief no doubt looks different from ours, but I can't doubt that they feel it. They're social animals with social bonds, and those connections have to mean something to them. Researchers have documented coyotes patrolling their territories in search of lost companions, pups howling and pacing when their mothers don't return. If they can grieve, I have to assume, too, that they carry the trauma and horror we've forced upon them.

―――――

The way we go after animals like coyotes, you would think we were being preyed on at every turn. But walking through the city, I don't fear wild animals; I fear other humans. Specifically, I fear men. There was the man who stopped me as I hurried through Cal Anderson Park in Seattle after dark to tell me I had dropped something—my smile, it turns out—then called me a bitch when I didn't appreciate his line. The guy driving past who stopped me as I was walking home here in Portland to ask if I had a husband. The countless guys hollering at me from cars and from across the street when I walked home in DC.

To be a queer woman in the city is to be reminded that I can be subjected to violence at any time. The guy who muttered, "That's not right, I should get some of that," as I held hands with my then-girlfriend on a date in DC. The teenagers who yelled "fags" at us while we walked down the street in my hometown. The man who reached out and grabbed my arm in the middle of the day on a busy sidewalk in Portland as Leslie and I walked past.

I don't have it nearly as bad as others. Women, trans people, queer people, people of color—all are subject to threats by other humans on the streets and in our homes. And the lack of legal recourse is infuriating. In 2019, a trans woman was assaulted in Portland, and police said she had probably just been drunk and fallen down. Twenty states still have no laws on the books prohibiting housing discrimination based on gender identity or sexual orientation. Halfway into 2021, at least thirty-five state bills had been introduced prohibiting gender-affirming healthcare for trans youth, while fifteen had been introduced barring trans people from using the bathroom or locker room that matches their gender. In seven states, trans students are not allowed to participate in sports teams for their gender; in Tennessee and Montana, it is now nearly impossible for students to learn about queer history or identities. On the street, in the workplace, at school, trans people are constantly under threat.

In 2020, at least forty-four trans and gender-nonconforming people were killed just for being who they are. Most of them were Black and Latinx trans women. If that's not predation, I don't know what is.

———

Coyotes, like many animals, get a bad rap. They're blamed for disappearing pets, slain sheep, and more. And yes, they do sometimes kill pets, whether for food or because the pets seem like potential territorial competitors—though cats would be better served by being kept inside where they can't get hit by cars and won't kill an estimated 2.4 billion birds each year. And it's true, sheep are bred to be docile, making them exceptionally easy prey.

But the fact of the matter is that coyotes help us more than they hurt. In cities and in farmland, most of what they eat is rodents—coyotes may be the ones keeping rats from going through your trash. By that token, we should be encouraging coyotes in our cities, not eradicating them.

We've just barely begun to study coyotes, too, rather than seeing them as pests. We don't even know what we would lose if we *did* manage to get rid of them all.

When settlers set out to eradicate coyotes, they also sought to rid the continent of wolves and other predators. With wolves, they largely succeeded. When Yellowstone National Park was established, for example, rangers worked diligently to kill the wolves that remained, and by the 1940s, wolf packs were rare in the park, if they existed at all.

In the absence of this apex predator, the ecosystem suffered. Elk herds ballooned, and the elk browsed heavily on cottonwood and aspen saplings, eating so many that no new trees could make it to adulthood. The forest essentially vanished in parts of the park, and with it went other forest-dwelling species.

In 1995, the National Park Service reintroduced wolves to Yellowstone. The ecosystem had been disrupted for more than fifty years. It remains to be seen if some species will return; some of the harm that came from exterminating the wolves may be irreparable. But so far, at least, now that they're back, the wolves seem to be helping to return some of the biodiversity the park was once known for.

Coyotes aren't typically apex predators; they hang out in the middle of the food web, picking off smaller prey and scavenging meals that wolves and other larger animals have brought down. But it would be absurd to assume they don't play an important role in ecosystems. Can we really afford to eradicate them when we don't have the slightest sense of what that might do?

———

One response to trauma is to create walls, to close off the part of us that was once wounded so that it can never be vulnerable again, cordoned off by a thick sheaf of scar tissue.

Perhaps our faith in walls is why we try to build our cities separate and distinct from nature, why we try to let in green space only in small patches that we can control, where we can create order through ornamental plants and carefully paved paths. It's so easy to think that if we dwell within our own shining bastions of civilization, nothing wild can ever hunt or harm us again. The more we imbue our homes with culture and artifice, the less power nature will have over us.

But as coyotes and other animals are proving time and time again, the membrane between city and nature is a highly permeable one. Urban predators remind us that humans aren't necessarily at the top of the food web. No police force, no neighborhood watch, could ever be enough to overcome the wilds represented by coyotes and their ilk.

———

When settlers showed up on the shores of this continent, they saw the land as their own, ready for the taking. Anything foreign in nature—an Indigenous community practicing a system other than agricultural, Christian capitalism; a canid ready to take advantage of domesticated flocks—was immediately deemed wrong. I wonder if any of the settlers ever realized that this was a choice, an arbitrary one at that: that they could have arrived and understood the Indigenous people as allies in the human world. They could have seen coyotes as essential parts of the local ecosystem. The newcomers could have asked where they could fit in rather than attempting to mold the world to their liking.

Instead, Indigenous Peoples, wolves, coyotes—all were left to justify their existence in the face of European invasion. They became "nature," wild and savage, while Europeans stood for all of culture, all that was right and noble.

Seeing nature and culture as wholly different and distinct is what got us into this mess of climate change in the first place. We tried to control the world around us, pulling ancient matter from the ground and burning it, eradicating entire chunks of ecosystems, without bothering to think about long-term ramifications—we just wanted to make things easier for ourselves for the next five minutes, the next year, the next decade. We forgot that we can't just take without consequences; we forgot that we're inextricably part of a larger system. No matter how separate we consider our cities from the "wilds" around us, we are indeed of this earth.

———

For the last fifteen years, the Indiana Coyote Rescue Center has been surrounded by a tall wooden fence.

Coyotes aren't very popular in rural Indiana. They're seen as a nuisance and as a threat to cattle, sheep, and other livestock, which many people depend on for their livelihoods. As a result, the rescue center is sometimes controversial. In 2005, a man came onto the property and shot one of the coyotes, Amber, in the face. Miraculously, she lived, though she lost her sight in one eye forever. Interns from the nearby Wolf Park, a wolf sanctuary and outreach center, came to the rescue center and tacked plywood over the chain-link fences to help keep the coyotes safe until a more permanent structure could be built.

So now, the coyotes keep to themselves, sheltered or cut off from the world around them, depending on how you look at it.

I wonder, though, what they dream of when they nap. I like to picture Artemis stalking the streets of Indianapolis, looking for rats and discarded meat that she can bring back to Orion and their pups. Neegan rolling around in a freshly fallen snowbank, scratching an itch that's taken a year to free himself of. And then there's Ares, on his way across a fallow cornfield, pausing before pouncing on a field mouse and catching it by the tail, a snack even better than a spoonful of peas.

PART TWO

"How can settler society, which possesses no fundamental ethical relationship to the land or its original people, imagine a future premised on justice?"

—Nick Estes, *Our History Is the Future*

5

EXTINCTION

On September 7, 1936, the last known thylacine died in captivity at the Hobart Zoo in Tasmania. He had outlived every known wild thylacine by six years, though probably he didn't know that. Perhaps he'd paced his enclosure wondering when he'd be reunited with his family and friends; perhaps he'd loped along believing he was only one thylacine among many across the island, cut off from the others by the fine lattice of a metal fence. Or maybe he knew he was alone, had known the end was coming the day he was captured, awaited it with a certain resigned dread.

Thylacines were also known as Tasmanian wolves. They looked a lot like dogs, though they weren't canines. Rather, they were marsupials: both males and females had pouches like kangaroos. They could hop, also like kangaroos, though mostly they walked. Europeans encountering them for the first time thought they might be a sort of hyena, or maybe a large—*very* large—opossum. They weighed between forty and seventy pounds, a bit more than a coyote.

Once upon a time, their range stretched throughout Australia and Tasmania. About two thousand years ago, their population declined precipitously in mainland Australia, though it's hard to say why. Perhaps dingoes, thought to have been introduced by seafarers from nearby Papua New Guinea or to have migrated via land bridge, competed for territory and prey; perhaps Indigenous humans pushed the thylacines out of their range. Or maybe both, or maybe something else entirely. No matter how it happened, by the mid-1800s, they were scarce on the mainland and more common in Tasmania.

From there, the thylacine's story is not unlike that of the coyote in the United States, though with a much more desolate ending. In 1830, Van Diemen's Land Company introduced thylacine bounties in Tasmania at $160 a head in today's equivalent, under the argument that they threatened sheep. They didn't—recent computer models suggest that the thylacine's jaw was too weak to allow it to hunt prey the size of sheep. But it was a compelling story, and between 1888 and 1909 the Tasmanian government got in on the bounty game. Hunting thylacines became a good way to make a living.

By the time everyone stopped paying for dead thylacines, it was already too late. When the thylacine was added to the Australian government's list of protected wildlife in 1936, it had been years since anyone had seen one in the wild.

That last captured one in the Hobart Zoo has become known as Benjamin. A brief black-and-white video clip of him remains. In it, he paces his cage, doing laps along the chain-link fence. He seems gaunt, but that could just be the thylacine's characteristic slenderness. His back is crossed with stripes not unlike a tiger's. He yawns, his long, thin mouth gaping like crocodile jaws. He pulls apart some meat. He lies in the sunlight restlessly before getting up.

In the old, choppy film, Benjamin looks almost animatronic, his motions jerky and awkward. His huge head is cartoonish. Someone coming across him for the first time could be forgiven for thinking he was a character in an amusement park or a Disney movie. The video doesn't bring Benjamin to life; rather, it's as if he was never real at all.

————

When I lived in DC, I'd go running most mornings before work. I'd cut across Columbia Heights and Mount Pleasant, then dip down the hill into the National Zoo. The zoo, part of the Smithsonian Institution, has no admission charge, and though it was technically closed until 8 a.m., its gates were never shut.

Depending on the time of year, I was either running in the dark or in the dim light of dawn. I'd jog up the hill of the Olmsted Walk, the wide pedestrian boulevard that cuts through the zoo, passing only the occasional other runner or zoo employee arriving for an early shift. It was quiet then, peaceful, with no traffic, no noise but the rare eldritch whinny of a waking zebra. I'd keep an eye out for the bison, sometimes dozing near the edge of their enclosure, but for the most part there wasn't much to see at that hour in that part of the zoo.

The way back down the hill was a different story. I'd retrace my steps down the Olmsted Walk, but halfway I would peel off toward the Elephant Outpost. The narrow trail there was hemmed in by trees, and in the winter it was fully dark except for a dim blue beacon over a safety telephone. My breath would quicken—what if one of the tigers had escaped? Or, more realistically, what if a wild deer were to crash through the bushes and straight into me? Once, a deer passed me on the Olmsted Walk, and my heart leapt into a sprint before I saw the deer for what it was and laughed at myself for being so skittish. I never encountered anyone, animal or otherwise, on the darkened cut-through.

The elephants were never out at that hour, though I always paused and looked, just in case. Then it was down the hill onto the American Trail. I'd stop briefly to see Crystal and Coby, the wolves, Crystal's white coat stark against the early dawn haze; I'd listen to their mournful morning howl. Then I'd continue, passing otters slumbering in an adorable heap and beavers tucked in somewhere under their dam. As a kid, I'd always envied Snow White and Sleeping Beauty for their ability to speak with forest animals, and my trot down the American Trail was a brief foray into a make-believe friendship with the animals across the barriers.

When I heard a few soft splashes, I was almost at my favorite spot.

The seals, a harbor seal and several gray seals, were almost always snoozing on rocks or bottling straight up in the water, resting with their heads above the surface while they gently spun. But the sea lions were up and at it every day. I'd stand at the edge of their pool and stretch, watching as they swam laps around one another, snuffling and huffing when they came up for air and then diving back below for another sleek swim across. I could have watched them for minutes, for hours, all day if I'd only had the time.

I spent those brief moments marveling at their acrobatic agility, knowing how lucky I was to visit them every morning. But also my heart hurt, knowing that this small pool could never measure up to the open seas their cousins swam.

———

My first encounters with wild animals, endangered and otherwise, were at the National Zoo. It was one of my favorite weekend excursions with my parents, a chance to see the lions, the cheetahs, the tigers, the elephants—and my favorites, the bats, fluttering around clumps of fruit in a darkened underground building. I'm less exuberant about zoos now, uncomfortable with the idea of caging animals that should be roaming free, but still, there's little doubt in my mind that the zoo was one of the places most responsible for my love of nature. It fed my certainty that I would be a biologist, or a veterinarian, or *anything* as long as I could work with animals.

In those visits during the '90s, the zoo still had its past stamped all over it: the enclosures were largely brutalist concrete structures; the Great Ape House was a loud, smelly building where plexiglass windows left the gorillas and orangutans few places to hide. It was an echo of the first zoos in the United States where steel-barred cages with concrete floors hardly gave animals room to range, let alone get some private time from jeering crowds. Animals had little opportunity to engage their minds or to practice the skills or habits they would have in the wild. They weren't thought to have emotions or feelings; there was no understanding at that time that by caging them, we might be traumatizing them.

Since those early zoos, things have changed. We have begun to understand that animals, too, have rich emotional and intellectual lives; accordingly, as their keepers, it is incumbent on us to make sure their lives are as pleasurable as possible. "Enclosures," not cages, house animals. No longer simply small concrete spaces, enclosures now mimic animals' environments and provide enrichment and entertainment. We get to go see the animals, but the animals have things to do, places to hide, and a choice as to whether they'll be seen. Zoos serve as a gateway into the natural world for children and adults alike. As research institutions, they are also crucial for protecting what wild animals still exist.

As a kid, I was obsessed with Tamora Pierce's novels, rich fantasy stories led by fierce, flawed women who found opportunities when none seemed to exist. There is a moment in one novel I read over and over again, *Emperor Mage,* where the main character, a young woman who has magic that enables her to talk with and shape-shift into animals, encounters animals living in a menagerie. She bristles at their captive fate and grieves for them. Drawing on her magic, she gives them all a waking dream, the ability to slip into a lifelike memory of the home they were stolen from.

The "enclosures" we keep animals in are a step toward that waking dream—but they are still cages. In calling them enclosures, we lie to ourselves, turning away from the essential fact that we are erecting a barrier between the animals and their homes—homes that in many cases, we have destroyed. I still love seeing animals, but each time I go to a zoo and pause to wonder at an animal, I grieve for it. I'm never quite sure if all this is worth it.

———

Zoos in the United States began to flourish at a time when Charles Darwin's theory of evolution was taking hold in the popular imagination. Though *On the Origin of Species* never explicitly grappled with the evolutionary link between humans and other animals, the implication that humans and animals are not so different was clear. It's easy to imagine that the spread of Darwin's ideas would have led to a shift where animals were treated more humanly—more humanely. And eventually, that shift would begin to happen, albeit far too slowly.

In the short term, American and European society grabbed on hard to the idea of competition and natural selection. Only those who were the strongest and the best would survive; all life was conflict and opposition. Although Darwin avoided mentioning human evolution, he drew clear lines between cultures and races, dividing them into "civilized" groups, "savage races of man," "half-civilized man," and "the lowest savages"—the last category the peoples of southern Africa and those descended from them.

Academics and others were quick to apply Darwin's ideas to humans, positing that those who did best—in finances, in health, in social rank—did so because they were better adapted and more fit than the others. Scholar Herbert Spencer coined the phrase "survival of the fittest," marrying evolutionary theory with racism to declare that "dominant races" were better equipped for modern life than "inferior races." Nonwhite humans weren't *animals*, according to these men, but they weren't exactly *Homo sapiens* either.

And so, at the turn of the twentieth century, several zoos began to display humans. In 1895 and 1896, the Cincinnati Zoo invited a group of Oceti Sakowin to live at the zoo for several months, participating in displays while they were there. In 1906, the Bronx Zoo displayed a Congolese man named Ota Benga in its primate exhibit, promoting him as the evolutionary "missing link" between humans and primates. These people were seen as less evolved, less human than the people outside their cages.

It's easy to think we've learned: we know now that all humans are human and that all humans deserve our care and consideration. But while we may not show off people of color in zoos anymore, we have a long way to go. We use Indigenous people as mascots for sports teams, allow our police to kill Black people with impunity, separate Central American children from their parents and literally keep them in cages without caretakers or access to basic hygiene when they and their families attempt to migrate to the United States, because they're somehow less deserving of these stolen lands than we white people are. We deny the humanity of people of color in TV and literature too: the very same book that I loved for its treatment of animals, *Emperor Mage,* treats people of color as uncivilized, cruel, and animalistic. And we condemn them to the worst effects of climate change: flooding, drought, heat waves, and more will be worst in low-income countries and communities of color. How can we claim we're any better than our past selves?

―――――

About six thousand animal species—from tiny butterflies to mammoth polar bears—live in accredited zoos and aquariums throughout the United

States and Canada. In some cases, zoos serve as the last-ditch effort to keep endangered species alive: they shelter vulnerable populations, encourage breeding, and when possible reintroduce animals to their wild environment. The National Zoo is working to breed the scimitar-horned oryx, an elegant antelope with long, curved horns, for reintroduction to the Sahara and Sahel where it is currently extinct, while the San Diego Zoo is attempting to grow the population of the 'alalā, or Hawaiian crow. Zoos have bred and successfully reintroduced populations of California condors, black-footed ferrets, golden lion tamarins, and other species all over the world.

But sometimes, even the most well-meaning captivity won't cut it.

Vaquitas are tiny porpoises, the tiniest of porpoises, measuring less than five feet long and weighing about a hundred pounds. I am bigger than a vaquita; unless you're a very small human or not yet finished growing, odds are that you are too. They're almost impossibly cute, with a blunt snout, dark panda-esque rings around their eyes, and a wee mouth upturned into a constant smile shape. Vaquitas also have the most limited range of any cetacean in the world: they live only in Mexican waters in the upper Gulf of California.

That range happens to coincide with the home waters of a fish the size of a football player known as the totoaba. From the 1920s into the 1940s, a commercial totoaba gillnet fishery thrived. Gillnets form a sort of wall in the water column, buoyed by floats at one end and sunk by weights at the other. The gaps in the net are sized for a certain kind of fish: the target fish's head can fit through, but its body can't, and as a result, when it collides with the net it's caught by its gills, left to struggle by its throat in vain until it dies of exhaustion or the net is finally raised. Gillnets have an incredibly high rate of bycatch—animals that aren't supposed to be caught but get stuck anyway—because they grab hold of anything generally the same size as the target fish. The damage they do ripples out through the ocean ecosystem.

Vaquitas happen to be just the right size for totoaba gillnets. As porpoises, they don't have gills, but they still get entangled. And because they breathe air, entanglement can be a death sentence—stuck in a net beneath the ocean surface, a vaquita will panic and drown.

We don't know just how many vaquitas perished in the totoaba fishery before it was closed in the 1970s, but the number was undoubtedly large. And though totoabas aren't officially fished anymore—they are still poached, thanks to a demand in China for their swim bladders—the waters of the Gulf of California are still flush with other fish. Gillnetting for sharks and rays has remained prevalent over the decades, and mackerel and shrimp trawls also entangle vaquitas. In 2017, the Mexican government outlawed the use of gillnets, but it was almost certainly too little too late. By the summer of 2018, at most, nineteen vaquitas remained, and the population continued to plummet.

In 2017, a group of experts from nine different countries led by Mexico and including the United States attempted to move as many vaquitas as possible into protective sanctuary in a last-ditch effort to keep the population alive. But it didn't go well. It turns out that vaquitas are stressed by being captured, and even more so by being moved. The first animal they caught had to be released when she showed signs of stress. The second animal they captured died not long after being placed in a protective pen.

To make matters worse, that second vaquita was a breeding-age female, one of the few who could continue the population. Not only did our interventions deplete the vaquita population by one, it also may have diminished the chance for future generations to survive.

We have, in essence, failed the vaquita twice: once by refusing to register the harm our pillage of the ocean did to them and so many other creatures, and then by continuing to slaughter them all the same.

Vaquitas have inhabited—and are leaving behind—a planet that has gone through more cycles and changes than we, in our short century-or-less lifespans, can truly comprehend. I have tried to picture the planet before vaquitas and thylacines and coyotes and whales and humans lived on it, when it was fully barren, and my mind comes up blank: even deserts teem with life, if you know where to look for it.

For the first couple hundred million years of Earth's tenure as a planet, nothing lived here, nothing at all. No plants, no algae, not a single living cell: the planet was busy stratifying matter from the cosmos into layers of crust and mantle and core and developing something that resembled an atmosphere. Everything, as far as the eye could see—if there had been eyes around to see, which there weren't—was mineral, elemental: rock and lava and pebbles and stone. Geology churned through the planet. Biology hadn't been invented yet.

Then, about two billion years ago, something magical happened. Deep in the ocean, at hydrothermal vents, minerals began to organize into molecules, then into proteins, and then into something that began to look like life. For their first billion years of life, these tiny protocells were just about the only living thing on the planet, burping along in the water column, eating one another and being eaten, slowly evolving into bacteria and archaea. Then unicellular organisms joined forces. The new multicellular organisms grew more complex, more fabulous, left the water—you more or less know the story from here. Life on Earth jumped to insects, to dinosaurs, to mammals, to—us. But along the way, there were five great extinction events, events that wiped out huge swaths of life and left what remained to reorganize itself into new systems.

The first was at the end of the Ordovician Period, about 450 million years ago. One theory says Earth suddenly got a lot colder. Creatures that were used to warmth froze. Glaciers formed and then melted, sucking up sea water and releasing it in cycles, and in response the sea level rapidly rose and fell, disrupting coastal ecosystems. Nearly 85 percent of marine species died, leaving those that survived to repopulate the seas.

Then, about 375 million years ago, at the end of the Devonian Period, trees evolved. Remember, for billions of years, this planet had no trees and no plants, just water and rocks and the occasional bacterial mat. Those new trees developed deep rooting systems and broke down rocks into the planet's first soil. Nutrients from that soil probably ran off into rivers and from there to the sea, choking massive, lush coral reefs. It was a total transformation for an ocean that had never really known land-based sediment.

Life regrew, though without roughly a quarter of the families of life that had existed before the Devonian extinction. Trees increased the oxygen concentration in the atmosphere, and bugs thrived, becoming bigger and bigger. *Meganeura,* a giant dragonfly-like insect with a wingspan of two and a half feet, flew through the air; a terrifyingly long eight-and-a-half-foot millipede relative crawled across the land.

Twenty-five million years later, things changed again as volcanoes erupted all across the planet. The change in atmospheric composition would have been something like what we're causing today, as volcanoes gave off huge quantities of carbon dioxide and other gases, trapping heat near the planet's surface. These eruptions also left dust clouds hanging in the air, blocking the sunlight plants needed to photosynthesize and turn carbon dioxide into oxygen. This extinction event—the Big One, the Great Dying—wiped out 96 percent of all marine species and 70 percent of terrestrial vertebrates. Among many other things, the Permian–Triassic extinction is why we no longer have trilobites.

Most of us are more familiar with the next two events: the Jurassic, about two hundred million years ago, and the Cretaceous, sixty-six million years back. The Jurassic is something of a mystery, taking out 75 percent of species—ammonites, marine reptiles, amphibians, and more—with no clear cause. But the Cretaceous–Paleogene event we know well: this is the one that exterminated the dinosaurs when an enormous asteroid hit Earth in what today is Mexico. The asteroid sent blistering vapor and debris all across the North American continent, incinerating everything in its path. It blasted pulverized rock into the air, where it briefly incandesced and superheated the planet's surface before blocking out sunlight and plunging the world into perpetual winter. Populations of dinosaurs, birds, mammals, and sea creatures were shattered. It would take millions of years for diversity on Earth to bounce back.

Near where I grew up, a series of steep clay cliffs rise above the Chesapeake Bay. They formed ten or twenty million years ago, after all of these extinction events had come and gone. Then, a shallow sea rested where Maryland is now. It was a home to sharks, whales, rays, all the things you might expect in the temperate Atlantic now, but also—because it was, then, much warmer—alligators and other reptiles. Over time, the world grew colder, the sea evaporated, and the cliffs were left behind.

These days, whenever it storms, the cliffs erode, leaving rubble of rocks and ancient fossils behind on the beach. The beach is now part of a state park, and fossil-hunting is welcome. I'd been there as a kid, though I hardly remember it, and when we lived in DC, I took Leslie back.

It was a warm June day, not quite warm enough to swim but perfect for looking through the mess of shells and stones strewn across the sand. The most recent early-summer thunderstorm had hit a week ago or more, and the beach seemed picked over. We could see fossilized shells packed into the clay edges of the cliffs, but on the beach there seemed to be nothing left for us but pebbles along the wrack line. I got bored and sat by the cliffs, looking out at the silty waters of the bay and trying to imagine it as it had been before white people arrived. Was it clear and blue, filtered by oysters and seagrasses? Or had it always been the uninspiring muddy brown of my childhood?

We were, as far as I could tell, the only queer people on the beach. It was mostly families with young children, there for a day trip to get their kids out of the house and maybe teach them some biology along the way. Toddlers skirted the waves that lapped gently at the sand, and preteens horsed around with footballs. The ecosystem here was familiar to me—I had known brackish Chesapeake water before I knew the ocean, knew the humid feeling of Maryland sunlight on my skin—but I couldn't help but feel a little out of place.

I watched as Leslie walked up and down the beach, their head down in search of some clue from the distant past. Something caught their eye and they crouched, brushing aside seaweed and driftwood to pick up something from the sand. Then they stood and walked over to me, elated. A small, dark gray object sat in the palm of their hand—a tooth, fossilized over time to stone. It was a thread pulling us past all these straight families and back to another world, one where this so-familiar beach would have looked like another continent entirely.

We found two teeth that day, and there were more, I'm sure, along that beach. At the trailhead there was a key for fossils commonly found at the cliffs, and we learned our teeth had belonged to saltwater crocodiles who had swum these waters some thirteen million years before. Lifetimes and lifetimes and lifetimes ago. They had died near where these cliffs were forming or had lost teeth to the hunt or to a fight for their lives. These bones were buried until a storm sometime in spring 2016, when they were uncovered and left to the Chesapeake. They were lost to time until two queers walking along the beach scooped them up and took them home.

———

Of course, you won't find saltwater crocodiles in Maryland anymore. The climate's all wrong for them. But their relatives remain in the reptiles found further south, in Florida and along the Gulf of Mexico. In that way, they live on.

This planet has survived more than us. Asteroids have hit it; volcanoes have erupted. Life on this planet of ours has been far more than decimated many times over, but each time it has rebounded, albeit radically different than before.

But while this planet has survived five great extinction periods, this sixth one we're in—where we're rapidly and profoundly changing the composition of the atmosphere, acidifying our ocean, and causing the death and destruction of species before we even know they exist—this extinction is the only extinction event in the history of Earth caused by a single species.

Let me state it more simply: this extinction is our fault. And if we don't do anything about it, sure, Earth will keep on keeping on. But we might not like where we end up.

Here is what is at risk if we keep going at our current pace:

One-third of all reef-building corals
One-third of all freshwater mollusks
One-third of all sharks and rays
One-quarter of all mammals
One-fifth of all reptiles
One-sixth of all birds

That doesn't even count the huge numbers of amphibians being wiped out by rapidly spreading fungal infections. It doesn't count bacteria and other microorganisms. It doesn't count deep-sea creatures we've never even seen before.

When we talk about extinction, we tend to look at a few large charismatic creatures: rhinoceroses, orangutans, tigers, polar bears. And it's true: these animals are truly, incredibly at risk. And they're big, fuzzy, often cute—we can relate to them. They feel almost human.

But the animals we focus on tend to be the ones that live far away from us: they're in the Arctic, or the African savannah, or the jungles of

Borneo. When we walk out our front doors, those of us here in the Lower 48, we don't see a polar bear clinging to a melting iceberg. We don't see an orangutan searching for her rainforest that was burned down to make way for cattle ranching or palm oil production. We can get in our cars and drive to work or off into the mountains where we won't notice the lupine species that have vanished or the frogs whose choruses we no longer hear. We don't have to notice. So we don't.

———

I have tried to get to know my local flora and fauna here in the Northwest. I have a local plant identification book and a tide pool guide, and I use the iNaturalist app religiously to learn what plants and bugs I'm looking at. Thanks to the pandemic, I've become a bird nerd. I know now, in ways I didn't before I moved here, how to recognize thimbleberries and trillium, fireweed and lupine. I know a fir from a spruce if I can get close enough to feel their needles; I know when the scat I see belonged to a coyote or a black bear, when a flash of red out of the corner of my eye is a northern flicker taking flight.

But I still can't recognize the wildflower penstemon to save my life, and I couldn't tell you the difference between a grand fir and a noble fir. A few summers ago, cow parsnip volunteered itself in my garden, and it took me all summer to figure out what it was. I am trying to learn, but it is taking time. I grew up only knowing the grass of suburban lawns, and clover, and buttercups, and dandelions—all of which had actually been introduced to the East Coast by European settlers.

In the days before globalization, people got to know their own local ecosystems intimately. They knew their local flora: what plants they could eat, what would heal them, what would hurt or kill them. They knew the rhythms of their seasons and the migration patterns of the animals they shared the land with. Many Indigenous people living in their homelands still know these things.

When settlers began relocating across the planet, many of us forgot our local knowledge. But we also began to see that ecosystems mirror one another; they create patterns.

Enter the biome. First introduced by Frederic E. Clements in 1817, biomes group species and habitats by common characteristics. You have your tundra, your coral reef, your forest. Though the specific species may vary within different iterations of the same biome, the archetypes are the same. If you go from one tundra to another, you will sense its essential tundraness. You will walk along springy, delicate moss that coats thick blocks of permafrost, ice-laden soil burrowed deep within the ground. You will run your hands along stones rough with lichen, pale greens and purples dotting the gray granite of erratics, huge stones left behind by receding glaciers. You will feel cold air whipping across your cheeks, carried swiftly across the treeless expanse.

At first, the categories of biomes were relatively simple and few. There were grasslands, deserts, tropical forests, and the like—broad terms that are in most people's vocabulary today. By 1961, biomes grew to include marine ecosystems, though only a small handful. The more we encountered varied ecosystems, the more we had to develop new terms for them.

By the 1990s, we finally began to realize we were in the midst of an extinction crisis—that thanks to human activity, animals, plants, microbes, fungi, entire ecosystems, all were going extinct at a much faster rate than was normal. We needed a more robust system. A group of researchers working with the World Wildlife Fund divided the world into 238 ecoregions within a series of "biogeographic realms." The ecoregions were roadmaps for conservation: at the very least, we should attempt to save *one* tundra, *one* rainforest, *one* desert. It's essentially a Noah's Ark of biomes.

But in all this, we forgot one key element: ourselves. Where do humans figure into the grassland, the rainforest? Does a grassland where there are farms operate the same way as a grassland left wild? Was there ever such a thing as a wild grassland in the first place? In 2008, scientists Erle Ellis and Navin Ramankutty asked the same questions and suggested that terrestrial biomes be reconfigured to consider human impacts. We're doing as much to shape our planet and its ecosystems as volcanoes, hurricanes, and other natural forces generally did before we showed up.

Traditional biomes like forest and grassland still exist. But in most places, they've been transformed: the forest has been logged for farmland or cleared for a new subdivision; the prairie has been grazed down by cattle and the bison shot, leaving invasive grasses eager to encroach. Amid these transformations, some original species or clusters of vegetation may exist, but they've become forest-and-lawns or grassland-and-pasture or tundra-and-oil-drill. So at this point, it's little more than wishful thinking to consider biomes without the human dimensions. Humans have altered an estimated 97 percent of land on the planet through habitat and species destruction. Each of us—individuals, governments—can't keep watching animals and biomes go extinct and pretend we have nothing to do with it.

———

In March 2018, I spent a morning helping two researchers band birds in a marsh in coastal Louisiana. Bird banding has been done in some form or another in the United States for more than two centuries; it's how we know about bird lifespans, migration patterns, and other habits. When you band a bird, you capture it, weigh it, record information about its sex, size, age, and other characteristics, and affix a tiny metal hoop to one of its legs. That band—so light the bird hardly notices it—has a number on it. If someone recaptures the bird, they'll be able to reference all the information the first researcher collected and can compare the present bird to its past self. Banding is, for example, how we know that Wisdom the Laysan albatross, the oldest known bird in the world, is more than seventy years old.

We caught birds using a mist net, which looks like a broad, finely filamented volleyball net that droops at each row. We set it up as quietly and subtly as possible in a part of the marsh that looked like a promising bird haven. Then we crept away from it, spread out, and marched back toward the net, flushing birds away from us and hoping a few might be caught in the net's threads, panicked but unharmed.

We were clumsy mist netters, most of us assistants total newbies. Our blundering steps through reeds and marsh grasses warned many of the birds away before we could even attempt to flush them into the net, and our poor coordination with one another let most of those who remained fly free above our heads. But we managed to catch a few, including one seaside sparrow, a petite brown bird with brilliant yellow patches over its eyes. I got to cradle it gently before letting it fly away, back to its life of hunting bugs and finding a mate.

The bird I held in Louisiana once had a cousin on the Florida Atlantic Coast. Walk through the marsh of Merritt Island before the mid-1900s and you would have heard a distinctive whistle-chirp followed by a buzz—the song of the dusky seaside sparrow.

Today, Merritt Island is home to the Kennedy Space Center. In 1962, as the space program kicked into high gear, NASA began to acquire land on the island. But the land, like much of the Florida coast, is swampy, and swampy land means mosquitos. To help deal with the mosquito problem and make the space center more attractive to engineers, scientists, astronauts, and other prospective employees, NASA flooded the island, covering the marsh in brackish water.

With the marshes went the dusky seaside sparrow's nesting land, now lost like some erstwhile Atlantis. The sparrow's population plummeted.

Around the same time, DDT was being deployed in the fight against mosquitos and other pests. DDT was, at the time, a groundbreakingly potent and effective insecticide. Its use is responsible for the fact that the United States has been malaria-free since 1949. But it's also extremely toxic, and one of its most notable effects in animals is thinning of bird eggshells, which decreases chicks' survival rate.

Between the flooding and the poison, the dusky seaside sparrow didn't stand a chance. The last one, named Orange Band, died of old age in captivity in 1987.

This little sparrow was lost to the world in the name of progress. Sacrificing it, we got to the moon; we eradicated malaria from North America. And the dusky seaside sparrow wasn't even its own species but a subspecies, *Ammodramus maritimus nigrescens.* Other seaside sparrows still exist.

Queer disabled writer Eli Claire points out that "life and death sometimes hangs on an acknowledgement of personhood." We've worked hard over the decades, the centuries, to convince ourselves that as humans, we are separate and distinct from others—other animals, other humans. It's made it easier to ignore the pain we're causing them and the damage we're leaving in our wake.

And when we don't feel or understand the pain we're causing, it's easy, too, to pretend that it's reversible. Because we paid people to kill them, thylacines were declared extinct in 1936, but we keep trying to convince ourselves we didn't do that much damage—they could still be around. More than a thousand errant sightings have been recorded since we admitted they were gone.

For a few decades, sure, maybe there was a chance that a few had snuck through and survived. But at this point, it's next to impossible. Still, people kept on looking. In 2021, the president of the Thylacine Awareness Group of Australia uploaded a video to YouTube that he claimed included pictures of a family of thylacines, and the internet erupted with hope. It was short-lived, however: within days, researchers debunked the photos as showing Tasmanian pademelons, small marsupials similar to wallabies. There went our chance for redemption, for the discovery that something we thought we had destroyed could be whole again. The thylacine is gone, and it's gone because of us.

We've destroyed so much, but so much else survives. When I held the dusky seaside sparrow's cousin in my hand in Louisiana, I could feel its nervous warmth, its fluttering heartbeat. It was alive and real, scared and breathing just like you and me.

6

RIVER

In the middle of a cold upstate New York winter in 1968, sparks from a welder working on a bridge set the oily Buffalo River in New York ablaze. A year and a half later, the Cuyahoga River in Cleveland, Ohio, caught on fire, marking what was at the very least its twelfth fire since European settlement.

These were relatively small affairs, fires put out quickly with little damage, but they marked a worrying uptick in flames where flames didn't belong, fire catching on oil slicks in the middle of flowing bodies of water. Then, on October 9, 1969, the River Rouge, a small river southwest of Detroit, showed them all up. A fire sparked not far from the interstate, the flames rapidly spreading some five hundred feet. They shot fifty feet into the air, bringing the river's traffic to a standstill as it blazed. It was a clear sign that something was seriously wrong.

Though those years marked a peak in riverine fires, the history went back much further: the Cuyahoga, for one, first burned in 1868. As settlers arrived in what is now the United States and established towns, then cities, they also created ever-increasing quantities of pollution. Sewage, industrial and agricultural runoff, and pesticides all wound up in our lakes and waters, causing blooms of harmful bacteria, poisoning fish, and, yes, catching fire. Then came the Green Revolution in the 1960s, and while expanded irrigation, advanced pesticides, and new fertilizers made it possible to feed more people than ever before, they also seemed to make our rivers even worse.

We knew that lakes and rivers throughout the world were dying, but we didn't know what, exactly, was killing them.

In response to calls from scientists and policymakers in the United States and Canada for research on pollution and freshwater ecosystems, Canada looked to northwestern Ontario. There, small lakes smatter the land like confetti. They are millennia old, scoured out by glaciers in the last ice age, essentially the puddles left behind when the ice receded. Because of their remoteness, the lakes were relatively unchanged by human activities, unlike the water closer to cities. So Canada established the Experimental Lakes Area, transforming about fifty lakes into sacrificial lambs.

First, researchers chose a few and began to add nutrients. They tweaked the ratio of carbon, nitrogen, and phosphorus to see which chemicals would encourage algae to bloom the most. When algae die, the bacteria that consumes them also consumes all the oxygen in the water. Eventually, so little oxygen is left that no animal life can survive. Essentially, the researchers were injecting the Green Revolution directly into the lakes and waiting to see what killed them. Then, the hope was, they would figure out what could fix them.

It worked: the experiments showed that phosphorus is the main driver of algae blooms and led to protective measures like reducing phosphorus in laundry detergent and similar products and removing it during sewage treatment so that it wouldn't poison our waterways. We killed a few lakes to help us not kill some others.

Over the decades, the lakes have been home to a whole host of studies. Researchers have tracked the effects of acid rain, altering the water chemistry to understand the impacts of burning coal and proving that the resulting increase in lake acidity is a death sentence for biodiversity. They've mimicked the flooding created by hydroelectric dams, which results in high levels of methylmercury when plant matter decomposes—and that dangerous chemical then bioaccumulates all the way up the food chain. More recent experiments have turned to the impacts of climate change, like how drought will impact the balance of life in fragile lakes.

The lakes have helped us understand our own cruel impacts on Earth. But they also might not have been necessary had we understood from the start that our actions would have indelible results—if we had looked for ways we could live *with* our planet instead of on it, against it.

It took millennia for these lakes to form, and just a few years to poison them. It may be centuries before they're back to normal.

———

In high school, I was a rower. I spent every afternoon on the Potomac River, working with my crewmates to find a humming unison that would drive our racing shell smoothly across the water. A few times a year, we'd practice in the early morning, and though I'd go through the rest of the day bleary and tired, those mornings were often my favorite. The river was perfect and calm, a glassy surface that stretched off into the pre-dawn quiet before the city around it awoke. As we paused and drifted between workout intervals, my mind would settle, and I'd listen to the river lapping against the shore, watch the mist coming off the water's surface. As we raced down the river, everything around us was so still that I could hear tiny bubbles rushing along the bottom of the boat. The river was refuge for me, a place of center and focus amid teenage turmoil.

But it was also something repulsive. Decades earlier, in the 1960s, the impacts of urban sewage and agricultural runoff had made the Potomac "disgraceful," in the words of President Lyndon B. Johnson. Cleanup efforts, luckily, had improved its conditions by the time I was rowing on it in the early 2000s. But still, when I flipped my boat one morning while rowing by myself, I showered as soon as I got back to shore, desperate to remove the chemical stench from my skin. And I remember the blade of my oar just barely missing the corpse of some animal during practice, probably an opossum or a raccoon, too bloated and rotting to be definitively identified.

The Potomac wasn't alone in its foul status. When we raced on the Schuylkill River in Philadelphia, other crews told us stories of rowing past the bodies of dead cows, and we weren't allowed the tradition of throwing our coxswains in after a victory. The Cooper River in Cherry Hill, New Jersey, had a certain unnatural odor to it. Even the idyllic pond at St. Andrew's School in Delaware, where we raced once a year, was clouded with thick weeds of *Elodea canadensis* that tugged our oars as they sliced into the water.

These waterways were our home for two hours every day, but they were neglected, ailing. It didn't take an expert to tell that something was seriously wrong.

––––––

It's been years since I was in a racing shell. But still, put me in a kayak or a canoe and I am happy, ready to paddle along a shoreline and search for turtles sunning themselves on logs, herons stalking fish in the shallows.

I don't actually like to be *in* the water, though. As a kid, the water terrified me, especially murky river water—what might be in there ready to nibble my toes, or worse? I'd jump in on hot summer days only to scramble out as quickly as possible.

As a ten-year-old, I spent a day trying to learn to sail on the Potomac River with my friends. Three of us crowded onto a Sunfish, a tiny, flat, single-sailed boat that a mildly skilled sailor can manage alone. The day was windless and hot, the sort of oppressive stillness that happens in the East Coast summer. With no breeze, we kept drifting and drifting, no matter how we steered or attempted to manage our sail.

The section of the Potomac we were sailing on was full of *Hydrilla verticillata*, an invasive aquatic grass that grows in long, green stems. Each time we drifted off course, we floated into the hydrilla forest. Within moments, greenery would be wrapped around our rudder and our keel, and we'd be stuck fast. And each time, as the strongest swimmer, I would have to jump in and push us out. The hydrilla would tickle my legs and grab at my arms, a mysterious sea monster lurking, ready to bite off a limb or carry me down to the bottom of the river. We were terrible sailors, and I lost count of just how many times I had to leap into the yawning maw of the weed-infested river. By mid-afternoon, I was in tears, and it would be years before I'd be willing to sail again.

————

The Potomac was one of the first North American rivers encountered by English settlers. Explorers made their way up the Chesapeake Bay and into its many tributaries, preparing to claim it all for themselves. In 1608, Captain John Smith noted beavers, otters, bears, martins, and minks. "In divers places," he and his men found abundant fish, "lying so thicke with their heads above the water" that he and his men tried to catch them (with no success) with a frying pan. It was an incredible bounty for a group of men who had likely eaten little more than weevilly hardtack for weeks.

The Potomac then was home to thousands of Algonquian people. Though the cultures of the region were largely agrarian, they depended on the river for the huge amounts of fish it held, and for the water it provided. It was clean enough to bathe in each day. The Potomac was a clear, biologically rich waterway, crucial for travel and for the Indigenous Peoples' way of life.

That wasn't to last long after the settlers arrived.

By the 1800s, the Potomac was in distress. The first sewer system began dumping waste from Washington into the river in 1810, and by the 1830s, fishing boats were pulling thousands of fish out of the river in a single haul. By the end of that century, the US Public Health Service reported that sometimes the river was so murky—and so full of fecal matter—that it was unfit for bathing, drinking, and cooking.

Through the 1900s, it got worse as the population surged and sent more sewage into the river. Mining in the West Virginia headlands sent toxins gushing into the mountain streams that fed the river. Agriculture all along the Potomac's banks flooded it with nitrogen, phosphorus, and sediment. Just a few centuries after John Smith's arrival, European settlers had turned the river—and so many others across the country—into a cesspool.

———

I, like so many other kids of my generation, first encountered John Smith in the Disney movie *Pocahontas*. Rivers are essential to the movie: in its opening sequence, Pocahontas swan dives past a cascading waterfall into a pristine blue river, while just minutes later John Smith and his men paddle across calm, clear, misty waters to take their first step in the New World.

Rivers are a metaphor for a philosophical conundrum too. "What I love most about rivers is you can't step in the same river twice," sings Pocahontas after her father, the Powhatan chief, asks her to make a pivotal decision about marriage. He has exhorted her to be like the river, "steady," but it's not steady at all, she claims.

The song—and the whole movie—is all about finding your own path and not just taking the easiest way forward. It's one of the main reasons

I loved it so much growing up. *Pocahontas* got me. Here was a girl who loved her family but also felt out of place, who seemed more connected to nature than to people. I pushed back against what people told me I ought to do—when I came out to my mom as queer years later, she would make a remark about how I always chose to do things the hard way. And I, too, felt more comfortable with animals than with people. I was happier in nature than anywhere else.

Watching the film now, I realize that some of the same things I connected with so much are what make the movie so problematic: it paints a portrait of an Indigenous woman who is more drawn to whiteness than to her own culture while also upholding stereotypes of Indigenous people as being magically, animalistically connected to the land—Pocahontas's grandmother is literally a tree who interprets dreams. And then there's the overt racism: in the first five minutes, the word "savage" is used to describe Indigenous people more times than I could count, and jokes about killing Indigenous people are slung around like nothing. Plus, the whole storyline is based on a falsehood: Pocahontas never fell in love with John Smith; she was kidnapped by white settlers, and it's more than a little likely that her subsequent marriage to Englishman John Rolfe was coerced. John Smith—voiced by none other than Mel Gibson, who has subsequently become famous for homophobic, antisemitic, and racist comments—is depicted in the film as sensitive and empathetic, but in truth, his writings show him as mistrustful of Indigenous people and more than a little self-aggrandizing.

I rewatched the movie to write this essay and I cringed all the way through, not sure how to reconcile my childhood connection with a narrative built so solidly on white supremacy.

There's part of me that wants to say that my connection to *Pocahontas* was because of my queerness, and because of my environmental-mindedness, and that's true. But I was also able to love the movie because it erased all that was so ugly about colonialism, or at least covered it up in an entertaining song, and I didn't have the knowledge to counter that narrative. There's nothing in the movie to indicate how the Powhatan, Pocahontas's people, were killed in droves by smallpox and other infectious diseases brought by the white settlers. Nothing to indicate that Pocahontas's sacrifice would be futile, as by the mid-1600s all Powhatan land would be claimed by white people. The river of Pocahontas's youth would be almost unrecognizable now, after four centuries of industrial onslaught. It certainly wouldn't be the same river she stepped in before.

———

The land throughout North America is carved and rearranged by rivers. This is especially true in coastal Louisiana, where change happens so quickly that no map is ever completely accurate. Coastal Louisiana is a vast alluvial plain, a pile of sediment that formed when the Mississippi and its neighbor, the Atchafalaya, changed their courses every thousand years or so and slithered back and forth across the floodplain like snakes, dropping mud as they shifted. That mud accumulated, forming land along the coast that developed into lush wetlands and depositing barrier islands further out that protected the coastline from storm activity. The river has been a builder, generating new life and new land where none was before.

Since the mid-1800s, though, we've restricted the Mississippi with levees and concrete channels, so it can no longer swing across the land. It's been part of an attempt to make the river's course regular and smooth, to support vast quantities of shipping containers and to keep the water from spilling over its banks and into riverside communities.

Constrained to one route, the river now requires frequent dredging to keep it navigable while land that would be rebuilt by a meandering river goes unreplenished. The silt that would normally be deposited all over Louisiana never makes it there. The Gulf of Mexico eats away at the wetlands, and the barrier islands disintegrate.

When I was in Terrebonne Parish, Louisiana—the same trip I spent banding birds—I took a boat out to Trinity Island, one of the barrier islands. The island is a small, low-lying, sandy hillock. It cuts in a stripe across the water, only a few paces across in width and about an hour's walk from end to end. Trinity is just a small relic of the barrier island's former self: once it was part of Isle Dernière, an island so large it held a resort town with more than a hundred summer homes. In 1856, a hurricane tore Isle Dernière apart, leveling almost every building and killing 200 people. Since then, the island has been disintegrating, so much so that it's now considered five separate islands, one of which is Trinity. Today, it provides a vanishing home to nesting birds and helps keep the Louisiana mainland from feeling the brunt of passing storms.

It took about an hour to get from the bayou to the barrier islands by boat. Gulls and terns wheeled and cried overhead while bottlenose dolphins hitched a ride in the boat's spray, disappearing in the muddy ocean before leaping miraculously to the surface once more.

When we got within sight of the islands, we slowed considerably, moving just barely faster than the current would naturally have taken us. The slope of the seafloor up to the islands is gradual, and all around them we risked running aground. When I asked the pilot why we were puttering so, so slowly, he pointed to the GPS. According to the computer, we were in the middle of the island. But here we were, floating in about three feet of salty, silty water.

The GPS wasn't wrong—it knew exactly where we were. Just months earlier, we *would* have been standing on the middle of the island. But each time it storms, more of the land washes away, and there's no river here left to replenish it. What might have been island just a week ago is ocean now.

About 40 percent of the wetlands in the United States are found in coastal Louisiana. It's hard to overstate just how important they are, to humans and animals alike. These swamps and marshes provide habitat for more than five million birds each year, and for fish, alligators, and oysters and other invertebrates. Wetlands soak up storm surge and prevent flooding, and feed the people who live near them. In coastal Louisiana, the bayou is the center of life: people talk about living "up the bayou" or "down the bayou" instead of north, or south, or the next town over.

Everywhere I went in Terrebonne Parish, the story was the same: the land was leaving, and the water was coming. People were constantly pointing out lakes and ponds and stretches of open water that had been marsh just a few years ago. They described the loss of their favorite fishing and shrimping spots, the disappearance of landmarks they once used to navigate, the flooding of protected bayou channels they once boated safely through. These ecosystems that people have depended on for generations are vanishing before their very eyes.

It's not just that the land isn't being replenished. As sea levels rise thanks to climate change, the land is more likely to get washed away. More frequent and stronger hurricanes also aren't helping: in 2005, hurricanes Katrina and Rita turned some 217 square miles of land and marsh into open water. As we continue warming the ocean, hurricanes will only get stronger and more frequent, and Louisiana's coastline will keep washing away.

Then there's the oil industry. As oil is sucked out of the earth, the land literally sinks, collapsing into the spaces left behind. Every well dug increases the likelihood that the land will fall beneath the water. To get to the oil and bring it more cheaply to port, oil companies constantly cut new channels in the bayou, increasing places the water can flow and erode the land. And as the 2005 Deepwater Horizon spill showed, even offshore oil spills can be devastating to the land. Marsh grasses and vegetation are still struggling to recover from the five million barrels of oil that spewed into the Gulf. Where vegetation has died off, erosion sets in, disintegrating what little marsh remains.

The shifting landscape affects everyone who lives there. You don't move to the Louisiana coast, generally speaking; you grow up there. Your family is there, your culture, your livelihood. And perhaps more than anyone, the inundation is affecting Indigenous Peoples who, with little representation on parish councils, are more disenfranchised than most when it comes to levee and flood management planning.

Two Indigenous tribes, part of the broader Houma Nation, live in Terrebonne Parish on the Louisiana coast: the Pointe-au-Chien and the Biloxi-Chitimacha-Choctaw of Isle de Jean Charles. Both tribes are amalgamations. Their ancestry goes back to the Indigenous Peoples of Louisiana who fled to the coast in the 1800s to escape religious tension, as well as to eastern Indigenous Peoples forced west by settlers. This mixed ancestry and migration means it's nearly impossible for them to gain federal recognition as a tribe, as the Bureau of Indian Affairs requires a tribe to trace its ancestry and connection to its homeland back generations. But after being displaced by the US government from their original homelands, they've lived here for years, and they're on the forefront of rising waters.

It's a dire situation. Since 1932, Terrebonne Parish has lost more than three hundred square miles of marsh and is on track to lose 40 percent of the parish over the next fifty years if nothing is done. The Pointe-au-Chien are lucky, by one way of looking at it: much of their land is within the state's levee system, which will keep some of the floodwaters out. But that's an imperfect solution—when I was there, tribal member Theresa Dardar explained that saltwater regularly comes into their land and makes it impossible to plant a garden, much less farm or raise livestock. And much of the tribe's traditional lands, including ancestral burial mounds, have been excluded from recent major levee plans. Before long, they may have washed away entirely. The tribe is "at the crossroads of adaptation or extinction," writes Pointe-au-Chien member and legal scholar Patty Ferguson-Bohnee—all because the state of Louisiana has deemed their lands not economically sustainable to protect.

The Isle de Jean Charles community may have to abandon their land completely. Their land is outside the state levees, and the only road onto the island regularly floods. In the early 2000s, the state considered including the island in a billion-dollar levee project known as the Morganza to the Gulf but decided not to because of the additional cost. Instead, they suggested the tribe relocate. A small ring levee built around the community failed in 2008, leaving the community flooded for days. In the past sixty years, the island has shrunk by some 98 percent. Within a few years, it will all be gone.

———

Just as coastal Louisiana is some of the most endangered land, the Colorado River is often considered the most endangered river in the world. For millions of years, it flowed along plains, over slickrock, and through the canyons of the American Southwest and into Mexico. After a journey of almost fifteen hundred miles, it filtered through dense wetlands and reached the ocean. At its full strength, the river's influence stretched for miles into the Gulf of California. In its silt, the river carried nutrients that fueled marine life from tiny clams to that cheerful porpoise, the vaquita.

Today, though, the river has changed. The population of the arid West has ballooned, starting with the gold rushes of the mid-1800s and marching forth into the tech booms of today. If you're in Las Vegas, Los Angeles, Phoenix, Tucson, San Diego, or many of the towns in between, you're one of the forty million people who depend on the Colorado for drinking water; if you eat California produce, you're consuming Colorado River water as well. We divert the river into reservoirs, dam it for energy, siphon it off for agriculture and our cities. We've extracted so much that when the river reaches the sea, it is little more than a muddy dribble—if it reaches it at all. The lush wetlands of the Colorado River Delta are ailing, where they still exist.

It's hard to understand what that really means, so let me restate. Once, so much water gushed from the river into the Gulf of California that it supported a network of wetlands two and a half times the size of Rhode Island. Now, some years, we consume so much of the river that not a single drop reaches the sea.

———

When the Colorado River began to flow across the Colorado Plateau about five million years ago, it was just a few streams of water seeking out the path of least resistance, the hollows between rocks, until they gathered in a braid of rivers. Over time, the water began to wear a groove, and the streams' coming together became a habit. The water returned to the same place year after year, and each year it wore down the earth a little more. It carried sediment downriver, and that sediment—sand and pebbles and rocks—scoured out the plateau like sandpaper. Then, funneled into the track that it built, the river became a torrent and raked rock and soil aside, eroding layer after layer of earth until finally, it lay at the base of steep cliffs.

No longer a meanderer after these millions of years, for nearly three hundred miles the Colorado River is constrained by walls a mile high: the Grand Canyon. Its walls tell a story that stretches back two billion years, when seas formed and evaporated in sequence across North America, leaving behind dirt, minerals, and the shells of tiny organisms. Over time, that all compressed into sandy, gritty rock. These rocks now make up the brilliant red and orange stripes that the canyon, and indeed much of the southwestern desert, is famous for. Run your hand over the stone and you can feel a rough memory of the world it came from.

Look closely at the layers of sandstone near the bottom of the canyon and you'll see fossilized stromatolites, layers of sediment built by the cyanobacteria that, through photosynthesis, helped add oxygen to the atmosphere and develop the breathable air we have today. Work your way up, and you're coming closer to the present day. In the Bright Angel Shale, 515 million years old, you'll find crinoids, deep-sea organisms that look like fans or flowers, relatives of which still survive today. In the 275-million-year-old Coconino Sandstone, you can see the fossilized tracks of reptiles and insects; at the canyon's upper reaches, in the Kaibab Limestone, you can find sponges and corals that lived in a shallow sea 270 million years ago.

It wasn't until the Colorado Plateau uplifted seventy million years ago that all this rock stretched above the sea. Over millions of years, the plateau rose, eventually reaching heights of eleven thousand feet, two miles of geologic history. Then, the river began to carve its route, layer by layer, peeling it all back. The Grand Canyon laid Earth's history bare, showing us the story of the continent we call home. Though it's easy to forget, water is stronger than rock; a movement is hard to stop.

———

Mojave poet Natalie Diaz describes the Colorado River this way:

> The Colorado River is the most endangered river in the United States— also, it is part of my body.

> I carry a river. It is who I am: 'Aha Makav. This is not a metaphor.

> When a Mojave says, *Inyech 'Aha Makavch ithuum,* we are saying our name. We are telling a story of our existence. *The river runs through the middle of my body.*

I spent hundreds, probably thousands, of hours on the Potomac River. I thought of it as home; I knew the stretch I rowed intimately. And yet, I always thought of it as something I was *on*, something separate. Its pollution was sad, a tragedy even, but at the end of the day it didn't really bother me.

But to say that is like saying one of my limbs was rotting away and I barely noticed or cared.

Diaz again: "The water we drink, like the air we breathe, is not a part of our body but is our body. What we do to one—to the body, to the water—we do to the other."

———

When Leslie and I moved from DC to Portland, we drove across the country and took a detour to see a few of the national parks in the Utah desert. We had been driving for days through brilliant reds and oranges, the colors of a sunbaked, parched landscape. The craggy Rocky Mountains had given way to high desert, vast flat land ringed by mesas that towered at the horizon. The deep gray-brown of granite had been replaced by sandstone that appeared almost purple in the dawn light but shone like gold at sundown. The blue sky felt impossibly large, endless.

All my life, I'd lived within the humidity of the Eastern Seaboard and the damp of the Northwest, and here I could feel my skin cracking as the desert sucked the moisture from my body. Water was scarce here, just muddy rivers and dry creek beds. A few times a year, the rivers would roar to life in powerful flash floods, but this was the dry season. Driving into Moab, we crossed the Colorado River, but hardly noticed—it was so thin and brown here that it all but blended into the slickrock.

During our stop to explore the parks, we stayed in an apartment in a subdivision on the outskirts of town. Its air conditioning was to be our refuge from the August desert heat. But when we pulled off the highway and turned into the subdivision, it felt like we had been transported to a completely different world.

Grass was everywhere—lush, green grass, thick and plentiful, not a spot of dried straw brown anywhere to be seen. Each house had the kind of excessively tended lawns I'd grown up with in the suburbs of DC, perfectly trimmed and edged. The deep, fertilized green was stark against the landscape's reds. It hurt my stomach to look at it.

There's a good three times as much lawn space as there is irrigated corn throughout the country. To maintain the average lawn, we pour in up to two hundred gallons of fresh, generally drinking-quality water per person per day. That's nearly ninety thousand Olympic swimming pools of water drained daily to make our grass grow, to make patches of the desert vibrant green. And all this costs money too: US spending on lawn care in 2015 alone outstripped the entire GDP of Iceland.

Desert lawns certainly aren't the only thing wrong with our water usage today. We pollute our rivers and lakes before we can even drink them, turning some of them so toxic that birds can't even land on them without dying. And most of the contents of our rivers—the Colorado included—gets diverted for agricultural use. But if we're willing to change the desert red to grassy green just to keep up with our neighbors, how carefully can we possibly be thinking about the river part of us?

———

When we moved to Portland, I had, for the first time in my adult life, a house and a yard that I could do anything I wanted with. For the past several years, I've been killing our lawn. I've painstakingly dug out sod to replace it with food crops—tomatoes, fava beans, snap peas, squash, garlic, a few failed attempts at corn—and I've collected cardboard and mulch to pave over the grass to create new soil, better soil, that I can replant with native plants.

I tell myself it's for environmental reasons. I don't want to water grass in the long Pacific Northwest summer droughts. I don't want to waste electricity or gas to keep it mown. I want pollinators to gather in our yard, so I can watch the hummingbirds and native bees buzzing from one flower to another.

And all of that is true. Since I've started transforming our yard, families of finches, bushtits, juncos, chickadees, and nuthatches have moved in; pollinators bumble around our flowers all summer long. But I also know that this isn't just an altruistic move on my part. I'm being selfish—I'm trying to distance myself from the perfect suburban lawns I grew up with.

Here I am, married, living in a single-family house that we own. There's a way of looking at my life that says I've bought into the system, that I'm little more than straight. So I yearn for something to make our home a little queerer. I tell myself that maybe this lawnless existence can be a tiny step.

————

Water is usually considered a renewable resource, part of an endless, perfect cycle. But while the total quantity of water on the planet may be fixed, that doesn't mean we don't change it by using it. After we're done, the water we have might not be the water we want.

We have hundreds of millions of square miles of water on the planet, but we're slowly turning it to poison as we fill it with pesticides and the chemical byproducts of mining, power production, and manufacturing. Just because the water was clean once doesn't mean it will be when we need it later. Our poisoned waterways destroy ecosystems and cause cancer and other diseases.

Human-caused climate change is also altering weather patterns: the water we've relied on in one area might not be there the next year or the year after that. California, the source of so much produce in the United States, saw unprecedented drought in the early 2000s. It was so dry that more than one hundred million trees died, putting the region's agriculture at risk and escalating the chances of wildfire. After a brief respite in 2019, the state is back in severe drought conditions. And while California was getting that momentary break in 2019, so much snow and rain fell in the Midwest that areas of North and South Dakota, Iowa, Nebraska, Minnesota, Wisconsin, Kansas, Missouri, Oklahoma, Arkansas, Tennessee, Illinois, Texas, Louisiana, and Mississippi all flooded. Four hundred counties sought federal disaster relief, and more than a million acres of cropland were ruined for the year's planting. As we warm the planet, this kind of water cycle disruption may become the new normal.

But water might be the catalyst that gets people to act on climate change. Oil extraction puts our water sources at risk. Pipeline spills threaten to contaminate drinking water sources and riparian ecosystems. Tar sands mining requires 2.3 gallons of water per gallon of oil and contaminates that water so it essentially can't be used again, maybe ever. Modern fracking events use an average of five million gallons of water to create and maintain wells and leave the water toxic, sometimes radioactive.

While carbon dioxide and other greenhouse gases are invisible and abstract, their effects nebulous and—we're convinced—far-off, water is life, and we know it. For years, Indigenous activists have been organizing around protecting their water from pollution and oil spills. "People who live with water still understand that Water is Life. The teaching is old, and it's profound," writes Anishinaabekwe activist and writer Winona LaDuke. "Our people have been protecting water since time immemorial, and we will continue to do so as long as we live." The rest of us are beginning to catch on.

While collectively we as a species may be unable to look far enough ahead to really believe the full impacts of climate change, most of us understand that to survive, we need to drink. "Wherever they live, people will fight for their water," says Naomi Klein—"even die for it."

In 2014, Dakota Access proposed a pipeline that would carry some 450 thousand barrels of crude oil per day from the Bakken oil fields in North Dakota to terminals in Illinois. Originally, the pipeline was planned to pass near the North Dakota city of Bismarck, but assessments suggested that the risks of leaks and other harm were too high for such a population center. Instead, the pipeline proposal was rerouted to just outside the Standing Rock Sioux Reservation. There, it would cross under Lake Oahe, a reservoir on the Missouri River that was created in the 1960s on two hundred thousand acres of the Standing Rock and Cheyenne River reservations without consent from or compensation to the tribes. The reservoir is no longer considered Standing Rock land by the United States and North Dakota governments, but it nonetheless serves as the tribe's main source of drinking water. What's more, the new pipeline route would potentially disturb sites sacred to the tribe. The tribe sued: the Army Corps of Engineers should have consulted with them but never did.

For months, the Standing Rock Sioux and tribal partners protested at the pipeline site. They camped out, gathering in opposition. And they were met with violence: sprayed with water cannons in sub-freezing temperatures, shot with rubber bullets. One burial ground was bulldozed by Dakota Access, despite the fact that permits were contested and awaiting review. Police from twenty-four counties and sixteen cities in ten different states joined local law enforcement to "manage" the protests with an army over thirteen hundred strong. A mercenary and security firm, TigerSwan, coordinated military-style counterterrorism methods on behalf of Dakota Access, comparing the water protectors to religious extremists and describing the protests as an insurgency in internal documents.

The rallying cry for the Standing Rock Sioux and their allies throughout all this has been *Mni Wiconi*—water is life. Rather than protestors, they are water protectors. In opposing Dakota Access, the tribe and its allies have been standing up to corporations and the United States government and asserting Indigenous sovereignty. They are also protecting a way of relating to Earth and its resources. Without healthy land and water, we have nothing.

————

The Standing Rock protests happened because the pipeline got rerouted from Bismarck—a city that in 2010 was 92.4 percent white—to just outside the Standing Rock Sioux Reservation. While an argument can be made—and in fact was made—that it was the larger population of Bismarck that caused the pipeline to be diverted, the role of privilege, too, is undeniable. The people of Bismarck had the legal means and the social standing to successfully reroute the pipeline. Indigenous Peoples like the Standing Rock, on the other hand, have endured generations of having their communities slaughtered, their families ripped apart, their cultures destroyed. Though tribes technically have sovereignty over their reservations, their ancestral lands often have been stripped from them and their protests systematically ignored or met with violence. Plus, they have little control over the land just outside their reservations, even when their drinking water source, like Lake Oahe, lies outside their borders, even when that land is theirs by treaty right.

And this is a pattern. The communities that have the least recourse, the most suppressed voices, tend to have their water most at risk. The city of Flint—which is majority Black—went without clean water for years due to neglect and a history of racist policies. Fracking threatens the water of rural populations in places like western Pennsylvania and the Dakotas. The tar sands in Canada and the Bakken oil fields, and pipelines carrying oil from them, threaten Indigenous treaty-protected hunting, fishing, and trapping grounds—and indeed have already destroyed many of these resources.

Water is a galvanizing force, yes, but whose water?

———

For the most part, I choose to identify as queer, rather than gay or bisexual. Like the river, queerness is fluid. Queerness offers me a subverting, a refusal to be dammed up and hemmed in by the levees of straight culture. It can be that utopia queer theorist José Esteban Muñoz talks about—"a horizon imbued with potentiality," one that seeks a world in which we navigate oppressions together, as a queer community, and return power to those who have been disenfranchised.

It is so tempting, as queers, to look to our future and let the past go, to move away from the scars and open wounds and attempt only to create our someday utopia. And it is tempting, looking at the state of our planet, to try to protect what we have left instead of thinking about what we could have had. It's tempting to pretend our queerness has nothing to do with the planet we live on, that queerness and the environment are separate, distinct.

But that leaves us empty-handed. To understand the river today, we have to look to the canyon it's left in its wake. To push against the system, to create something new, we have to look at what we've done, and we have to understand what has been done—especially what *we* have done, we white queers—to other people. To heal the land, we have to understand when and why we've claimed it as our own, and how we've neglected it. We can't make things better for ourselves, for others, for our planet, if we forget that the river is our body.

7

FIRE

In the Pacific Northwest, 2017 was a year of extremes. Instead of the usual fine, misty rain, the winter brought record snowfall, covering Wy'east in a glistening coat of white and dumping a foot of snow and ice on Portland's unprepared city streets. But then, summer ushered in oddly hot temperatures, the city hitting over one hundred degrees one day in June, normally a cool, rainy month. By late summer, the mountains were crispy. The bark and pine needles littering forest floors were desiccated, meadows brittle and bleached, creek beds empty but for tiny dribbles running down their centers.

When a teenager and his friends set off firecrackers on the Columbia River Gorge's Eagle Creek Trail, there was only one way for the forest to go—up in flames.

And so it did. Overnight, the firecracker-sparked fire flashed across more than four square miles of some of the most beloved public lands in the Portland area. It cut off the Eagle Creek trailhead, trapping 140 hikers who had been out for a simple summer day hike. Stranded with limited supplies, the hikers were forced to shelter in place overnight, not knowing just how close the fire would get, before evacuating to safety the following morning. All night and into the morning, Portlanders followed the story, terrified the hikers might not make it out and that our outdoor refuge would be destroyed.

The next day, the wildfire continued to spread. Flames encroached on the town of Cascade Locks, threatening thousands of homes and forcing people to flee for safety. It seemed inevitable that it would burn the historic Multnomah Falls Lodge to a crisp. The region watched in horror as the fire jumped the river into Washington.

A few days in, the Eagle Creek Fire merged with the nearby Indian Creek Fire, a low-intensity fire that had been burning—mostly under the Forest Service's control—since June. Together, the two fires grew hotter and spread over more than seventy-five square miles. Smoke and ash blanketed Portland. Some residents had to sweep thick layers of sediment off their cars. The sun disappeared and the air was so thick with particulate that it wasn't safe to be outside. I-84, one of two main routes into the city, was closed for days, the tarmac scorched and covered in debris.

It was the largest fire near Portland in decades, and it was, at the time, completely unexpected. But now, it seems to have been a harbinger of a new normal in the Pacific Northwest.

———

The summer of the Gorge fire was my first full summer living in Portland. I'd made my way there with Leslie the August before, driving across the country together from DC to a place I knew in my gut would feel more like home. The Northwest was a place of forests, of mountains, of queer community—everything I craved. I was ready to spend every glorious sunny summer weekend hiking. While I knew fire was a possibility, it was a distant one, just a matter of everyone being careful about their campfires.

But the teenager had lit firecrackers during a burn ban, and the hiker who witnessed his actions described him and his friends laughing as they tossed the fireworks into the tinder-dry trees below. It felt like they had committed a murder, their actions callous, neglectful, and cruel.

Portlanders were one step away from an angry, pitchfork-wielding mob, calling for the kids' heads. We assumed the worst—that the Gorge we loved would be blackened and dead, that our treasured trails would slough off the side of the cliffs in tremendous landslides. Leslie and I felt an outpouring of communal grief as we passed other hikers on coastal trails, the air hazy and heavy with smoke even so many miles from the fire. I heard more than one person describe themselves as heartbroken.

And really, heartbroken is exactly how I felt. I cried long, hot tears when I read the news, shook with rage when I called home and tried to explain to my mom how it felt to have my world on fire. I spent hours refreshing wildfire maps, hoping and hoping that the internet would tell me the fire was suddenly, miraculously contained. If I just spent enough time watching, then maybe, just maybe, some of my energy would flow to the firefighters and help them control the fire. Maybe the next time I refreshed the page, it would say something different.

One of the first hikes I'd done the summer we moved to Portland was Angel's Rest, a popular trail that ascends steeply to the top of the Gorge, where sweeping views of the river serve as a reward. Leslie had chosen it, a trail they'd done years earlier and now wanted to share with me. We'd encouraged each other all the way up the switchbacks, breathing heavily as the trail steepened and marveling at how easily twelve-year-old Goose made her way over the rocks. We'd hugged and kissed in triumph when we made it to the top, took in the glorious view of our home, the river wending its way through this miracle valley.

Now, Angel's Rest was on the edge of the Gorge fire. I wondered if it would ever be the same.

———

When I was twelve or thirteen years old, I started feeling sad and numb and quiet all the time. At school, I'd sit in the cafeteria with friends during lunch and put my head down on my binder and close my eyes, overwhelmed and antisocial. At home, I'd stay up late into the night chatting with strangers on the internet, looking for answers to questions I couldn't pinpoint. When my mom would kick me off the computer at one or two in the morning, I'd lie awake in bed on my back, waiting for morning, then stagger blankly through the next day, sleep-deprived and miserable. For a brief time, I cut tight rows into my wrists and ankles and upper arms, desperate to feel anything in a way I could understand, leaving light scars like tree rings etched into my skin. Eventually, I decided it was easier not to feel at all.

And when I was twelve or thirteen years old, I knew I was queer. I wasn't ashamed of it, not really, but I also didn't really understand what I was feeling. I told a few close friends I was bisexual, but I didn't actually date girls, didn't tell my parents or my brothers. I didn't know how a person would actually go about *being* bisexual: I didn't have a queer community, didn't know anyone who was really, truly out. In high school, I heard whispers of two girls I knew who might have been together, but then again, I thought, maybe that was just a rumor.

While no one in my liberal community ever said anything was *wrong* with being queer, no one really ever said anything at all about it one way or the other. We just didn't talk about it.

I didn't know anything about queer history, about the closet or the AIDS crisis or the ostracization of gay and trans people as mentally ill, but I think some part of me felt it in my body, a constant flickering flame of sadness that accompanied me everywhere.

There's an iconic photo from 1993 of the San Francisco Gay Men's Chorus. Five rows of men stand on choir risers, with about twenty men in each row. Almost all the men are wearing black and facing away from the camera, while seven men, scattered throughout the group, wear white shirts and face forward. Those seven in white are the only surviving members of the original choir; the men in black stood in for all those who had been lost to HIV. Seven men among more than a hundred, the only ones able to carry the choir's original legacy forward, the only keepers of the original memories. That same photo could have been taken in practically any queer community anywhere at that time.

When I was in college, finally truly, broadly out of the closet, I was obsessed with the AIDS epidemic. I wanted to know everything about it, wanted to understand who fought back, who took care of us, who failed us, who died, who we lost. I read obsessively, checking every book out of the library on the topic I could find. I watched the ACT UP Oral History Project videos, wrote term papers on the epidemic. I had read so much by my junior year that one of my professors asked me to co-teach a class session on the graphics used by ACT UP and other organizations to shock people into action.

I was grateful, of course, to be coming out in a time where my friends weren't all dying, to be living in a time where I was relatively safe from disease and so were my friends. I was grateful to be queer when it wasn't a death sentence. But also, in a bizarre way, I wished I could go back in time. I wanted to know everyone who had lived through the height of the AIDS crisis. I wanted to know the people we had lost. I carried around their empty memories like formless ghosts in my bare hands, and I wanted to give them shape, to feel them with me like the family they were.

———

I am drawn to trees, too, like family. Each time I hike, I have a tendency to find at least one tree that I need to stop and say hello to. It's often the biggest tree, one that has lived longer and seen far more than I can ever hope to live and see. But not always—once, I stopped for a slender sapling growing out of the decaying stump of an ancient tree. Leslie knows to expect it, the pause mid-stride as I cut off the trail just a few feet. I talk to the tree, thank it for being it, not that it needs my thanks. I put my hands on it to feel its rough, layered bark, inhale to smell its sweet and spicy evergreen scent. I love Douglas firs especially. As they age, their bark splits and furrows, forming thick scales as big as your hand, a small greeting from this towering plant.

On my first trip to the Olympic Peninsula, the year after I'd finished college and moved to Seattle, I drove to the Hoh Rainforest. The forest is, according to the National Park Service, "one of the finest remaining examples of temperate rainforest in the United States," which is to say, unlike much of the Olympic Peninsula, it has never been logged. It's still the old-growth forest that was there when it was inhabited only by the people of the Hoh and Quinault Nations, before European logging interests arrived.

I'd known in an abstract sense that trees could get unimaginably big, but I'd never seen anything bigger than the large tulip poplars that grew in my parents' backyard. It was pouring, and the drive took far longer than I'd expected, and I hardly wanted to get out of the car or take a hike, but finally I made my way onto a trail not far from the road.

Only a few strides off the forest road, stumps of trees that had fallen long before I was born stood strong, still decaying into the forest floor. They were so tall that it seemed that they must've kept growing after their trunks had cracked and their branches tumbled to the ground. Some stumps were at least twice as tall as me, and when I tried to put my arms around them, they were so wide that my elbows hardly bent at all. Even in death, decomposing, they were giants.

They were still giving to the forest too. From the center of some stumps, new trees grew, receiving the nutrients the old trees had spent long lifetimes accumulating. A single Sitka spruce in the Hoh Rainforest can support more than sixty different moss and lichen species, and the stumps were no exception: they were blanketed thickly in lush, deep moss that sprang back to the touch, the spongy greenery calling to my fingertips.

Even in the pouring rain, I was so grateful to be among those trees. I was falling in love. I marveled at their beauty while I mourned and apologized that we had cut their nearby cousins down, wondered how many other trees like them there would be if we settlers had just left this peninsula alone.

———

AIDS hit the gay community swiftly in the '80s: by March 1987, thirty-two thousand cases had been reported in the United States, and who knows how many others had gone undiagnosed or unspoken. But although the disease's spread was in epidemic proportions, local and federal governments were painfully, cruelly slow to respond. New York City mayor Ed Koch limited the amount of funding that could go to AIDS care and prevention. The Catholic Church, a huge source of social services in the city, refused to support education about condom use even in city-funded AIDS healthcare facilities, and actively opposed gay rights. Though the first cases of the disease had been identified by the early '80s, President Ronald Reagan didn't even say the word AIDS in public until 1985; his press secretary cracked homophobic jokes about AIDS multiple times when asked if the president had any plans to address the epidemic.

In the face of such apathy and derision, the only way queer communities and communities of color could get any help was to make noise—a lot of it. Throughout the '80s and early '90s, ACT UP and other groups organized attention-getting protests and die-ins to push for universal healthcare, funding for drug research and distribution, community-based education, and an end to discrimination against those living with AIDS. Their actions were also generally accompanied by graphic, often controversial posters and flyers designed to garner attention, gain support, and offend and shame people into action.

There was the Silence=Death poster, which inverted the pink triangle homosexuals were forced to wear in Nazi Germany and placed it on a large black field above the words "SILENCE=DEATH" written in stark white, framing the social inaction on the AIDS crisis as yet another form of genocide. Other posters depicted queer couples to advertise kiss-in protests, while a bus ad that said, "Kissing Doesn't Kill: Greed and Indifference Do," showed queer couples kissing—some of the first blatant and public images of queer couples showing any form of passion or affection I'd ever seen. And then there were the rage-filled posters placing the blame squarely where it belonged, on the public figures who failed to act: "He kills me" in red letters over a portrait of Ronald Reagan; "AIDS profiteer" over the face of the chairman of Burroughs Wellcome, which charged ludicrous prices for AZT.

"I am out, therefore I am," said a sticker based on Barbara Kruger's "I shop, therefore I am." The AIDS activism graphics were statements of pride even in the face of grief, of a desperate need to be seen—or else. I wanted to be part of that community that had grown together and refused silence in order to protect their own.

———

One of my favorite science words is *ecotone*. I love the roundness of the word, the fullness of it, and I also love its meaning—a gradient, of sorts, between different habitats. Another way of putting it is an edge habitat, the liminal boundary between one place and another. The tidal zone, where waves wash up onshore, covering and uncovering rocks and beach, is an ecotone. Marshland, the ribbon between river and land, is another. And the place at the edge of a burn zone, where some trees are scorched and others still stand tall, is a third.

Beyond the melodious sound of the word itself, ecotones have a fluttering beauty. They combine the best parts of different worlds, allow unexpected combinations to arise. In the liminal ecotone, more than one thing can be true.

For animals, an ecotone can provide safety and opportunity. At the edge of burn areas, old undergrowth is cleared away and the tree canopy is diminished, making way for sunlight and mountain breezes. New plants take hold, verdant shoots ready to be eaten by small creatures. When predators like coyotes or eagles come, grazers are close enough to the refuge of the forest that they can escape and shelter among the trees.

In 2011, fire burned across Mother Lode Mountain in Mt. Hood National Forest. Seven years later, I spent a few nights camping on the edge of the burn zone with a group of other writers, hiking into the burn area during the day to see how it had changed. I had expected desolate, empty land scarred by fire, but saw something else entirely.

Thick forest of Douglas fir gave way to open meadows crossing the mountainside. The lower trunks of trees that had been burnt still stood perhaps fifteen feet tall, monuments to what they had once been. Where the fire had touched them, the trees were a deep, glistening charcoal black that beckoned for me to rub my fingers on them. Mushrooms like tiny clamshells had nestled into bark ridges, slowly feeding on the wood inside. Other trees had shed their bark and their trunks had weathered to a shining silver, skeletons so luminous that you could be forgiven for thinking they were actually metal. In the distance, a woodpecker hammered away, searching for a meal.

Around the trees, plants were thriving—all that burned matter had fertilized the soil, making way for new growth. Huckleberries beckoned, and I picked them trailside, savoring their sweet–tart juice. Fireweed flowered in hot pink sprays, their tufts of seeds like dandelion heads waiting to catch on the wind. Rhododendron, Oregon grape, and salal shrubs sprouted up along the trail, while clumps of beargrass grew wherever they could find a foothold.

On one hike, I stopped halfway up the mountain, well up into the burn zone, and sat to have a snack and take in the view. The ridge across the way was burnt in a mosaic pattern, a patchwork of older uncharred growth and open meadows where the fire had done its clearing away. I wished I could jump across the valley, to sit and wonder at all the newness sprouting up around me all along the edges.

———

A few weeks before the Eagle Creek Fire ignited, Leslie and I stayed in an old fire lookout at Gold Butte in Willamette National Forest. We drove for a long time on dirt roads so rutted out that we moved at a crawl, the car dropping and rising in and out of huge potholes like we were on a rickety amusement park ride. Finally, the road ended and we unloaded our gear, trudging a mile up a steep, rugged path to the top of Gold Butte.

The lookout was a small, one-room shelter with windows instead of walls, perched on top of a rocky outcropping. A wooden porch wrapped around all four sides of the hut. Inside, there was just enough space for a twin bed, a small desk, a wood stove, and a little shelf in the center of the room. An old metal phone was still there, though when we picked it up, we heard only silence.

In the middle of the last century, when the lookout was active, the windows and the porch would have been crucial, used by a fire spotter during the long hot summer days and crisp starry nights to keep a lookout for plumes of smoke that signaled a disaster waiting to happen. But now, the windows just let in clear mountain light and let us take in the views before we got out of bed to meet the early morning chill.

To the north, Mt. Hood showed off its still-snowcapped peak, while Mount Jefferson loomed large to the east. I spent long hours exploring along the outcropping, Goose trailing behind me. We watched the light shift from dawn purple to daylight blue and then back to the particular golden afternoon light that happens best on mountain peaks. It felt like I could see everything.

On our second morning there, I stepped out of the cabin before Leslie was awake and settled onto the porch to take in the view with a cup of tea. The air had gotten hazy and an odd cloud had formed on Jefferson's flank. It rose from the forest floor and into the sky above. I watched it for a while, curious.

After long minutes, it dawned on me that I was seeing a plume of smoke. It seemed unreal—I knew forest fires happened out here, but I'd never actually *seen* one. After about an hour, once Leslie was awake, we remembered we were at a fire lookout, albeit an old and retired one, and I called it in on my cell.

The Forest Service had known about it, so my call, though satisfying, was hardly useful. There were probably people on the ground battling the fire as we watched from afar. We were miles away, safe, but still close enough to feel its presence in the air and in our lungs. Already, hot winds were fueling the fire, and after nightfall we could see flames shooting up into the air, a beacon against the starry sky.

———

The summer of the Gorge fire, Leslie, Goose, and I spent almost every weekend hiking. We would make our way up mountain trails, distracting ourselves from our tired bodies with stories and jokes until we reached a crest or peak. Then we'd rest and eat lunch, taking in the view of the Pacific Northwest stretched out before us.

Once the fire started, we had to be more particular about which trails we chose. We spent one Saturday morning driving to a trailhead only to learn, five miles from the parking lot, that the trail was closed to everyone but firefighters using it as an access point. The weekends after that, we went north to Mount St. Helens or west to the coast in attempts to find our way out of the smoke.

On every one of those hikes, our conversations always seemed to lead back to the kid who had started the fire. Everyone we knew was vitriolic. Seattle alt-weekly *The Stranger* referred to the "idiot teen" who had set the fire, and the teen and his family faced so much outrage and scorn that the state decided not to publicly release his identity. It must've been scary to be him, but we also didn't really care. He deserved it.

The heat from our own anger kept Leslie and me warm as the weather turned to fall and the mountain air grew cooler. We wanted the kid to pay. We'd lived half a year under the Trump administration, and it felt like everywhere we turned, men were fucking the world up. Already, the administration had attacked healthcare, withdrawn from the Paris Climate Agreement, attempted to block Muslim people from immigrating to the United States, and started promoting Trump's ridiculous border wall—and more and more and more, an endless onslaught of horrible plot twists. Through that whole spring and summer, trails had been our haven, where we could escape for a few hours at a time, where cell service wouldn't reach and we could pretend the world was normal and safe.

And then, a boy set it on fire, all in the pursuit of blowing some shit up for fun. Here was yet another example of masculinity at its worst, blindly acting without any regard for anyone else—human or otherwise. Here was proof. Even nature wasn't safe from men.

But had we ever thought it was safe in the first place? It hadn't really ever been, not for centuries; white men had been extending greedy hands over the land since they first arrived. How could we expect anything different just because we loved the place?

———

The Forest Service started because a few men wanted to undo some of the damage the United States had caused. As railroads expanded westward, businessmen sacrificed great swaths of forest to profit by way of mining and logging. Logging was a lucrative, destructive business, honed in Maine, Michigan, and the Appalachian Mountains and then exported to the West. The Forest Service, the brainchild of Teddy Roosevelt and Gifford Pinchot, was meant to save the remaining forests of the continental United States.

Pinchot, who would become the first chief of the Forest Service, was no big fan of fire. "Of all the foes which attack the woodlands of North America," he wrote, "no other is so terrible as fire." Still, in the early days of the Service, wildfire suppression was only a small part of the agency's work; most of the new rangers had their hands full trying to identify and save tracts of land from clearcutting.

But in 1910, five years after the birth of the Forest Service and at the height of an especially hot and dry summer, wildfires across the Bitterroot Mountains of Idaho and Montana merged and multiplied. Understaffed and underfunded, the Forest Service had almost no one to fight the fire, and no money to pay what few firefighters showed up. Within days, it had become one of the worst natural disasters the United States had ever faced. The Big Burn, as it would later be named, burned nearly five thousand square miles—an area sixty times the size of the Gorge fire—to a crisp. More than eighty people were killed, many of them firefighters who had taken the job out of desperation. Several towns were burned to the ground.

Fire was not new to the Bitterroot Mountains. Before settlers arrived, the land had been inhabited by the Schitsu'umsh, also known as the Coeur d'Alene, and other communities. The Indigenous Peoples of the region had purposely set fires to corral game and to renew the land so key plant species would grow. These controlled burns would have consumed some of the fuel drying out in the vast mountain forests and would have created a patchwork of burnt and unburnt land that could prevent huge fires from building up.

But as white settlers moved west in search of their fortunes, the Schitsu'umsh and their neighbors were forced off their land. With them went generations of fire traditions, and the stage was set for a conflagration unlike any the settlers had seen before.

After the fire, Pinchot changed his agency's priorities. Forest fires were preventable, he insisted, and what's more, had they had enough money and firefighters, the disaster of the Big Burn could have been averted. By 1935, his outlook was set in stone. It was known as the ten o'clock rule: any fire spotted anywhere in public forest land would be under the control of the Forest Service by 10:00 a.m. the next day. Fire became a war to be won, a foe to vanquish. Like so much else in this world, we thought we could control it.

When you stop a wildfire, you leave all its fuel unconsumed. That fuel doesn't go anywhere—it just waits to help stoke the next fire. It may look, for a short amount of time, like we've controlled it, but it's just a temporary illusion.

I was scared for years, as a kid, to tell people I was queer. But the more you tell someone to hide themselves, the more they want to show it. Now, I can't imagine putting that part of myself away.

As a teenager, I kayaked across part of Yellowstone Lake. I was part of a group, a summer camp of sorts, and we'd spent a week backpacking and climbing throughout Wyoming before heading to Yellowstone National Park. I'd never known a lake could be so big. Yellowstone Lake is about six times the size of Manhattan and sits in an ancient caldera. Strong winds whipping across the water can make paddling from one campsite to another feel far more like you're at sea than on an inland lake.

On a mountain ridge across the lake, the dark green of living lodgepole pines gave way to a dull gray brown. Charred and bare tree trunks stood like sentinels before us. Much of the burn area was a remnant of the infamous 1988 Yellowstone fires, the land still stark and scarred more than a decade later. That year had been unusually dry, with little snow falling during winter and still less rain in the spring and early summer. For decades, fire had been suppressed in and around the park, leaving the forest floor cluttered. Leaves and dead wood littered the floor; undergrowth became thick and tangled. Many of the lodgepole pines in the park were reaching the end of their lifespan, ready to die.

Early that summer, fires began to spark throughout the region, some from lightning strikes and others from human mishaps. As usual, personnel from the park and the surrounding national forests observed the ten o'clock rule, deploying firefighters to contain the fires as soon as possible.

But this time, it was simply too much. Fifty-one separate fires raged within the park, nine of which had been caused by humans and the rest by lightning. By late July, seventeen thousand acres of the park were on fire. The flames moved rapidly, sometimes advancing five to ten miles a day, moving even at night while humidity stayed abnormally low. Sparks from the main fires ignited new blazes; fires combined. Falling trees made it dangerous for firefighters to get close enough to suppress the spread.

All told, that summer was the largest firefighting effort in the United States up to that point—larger even than the efforts to stop the Big Burn. But it wasn't until the autumn rains and snows began that the fires stopped advancing and eventually sputtered out.

And then, a curious thing happened. The burned forest in Yellowstone didn't die; in many places quite the opposite. It thrived. Meadows grew in the burn areas, attracting deer, elk, bison, and smaller grazers. New lodgepole seedlings took root, sprouting new forests. It was a changed landscape, certainly, but flourishing in its own way.

In the years leading up to the conflagration, land managers had started to recognize fire's role in ecosystems and question the rule of suppressing every single fire. It was a rediscovery of sorts: the Apsáalooke (Crow) of the region had traditionally used fire to create meadows that would attract game, a practice that had ended with the creation of the park. But still, like the Forest Service's stance, the park's fire policy had continued to emphasize suppression. Only a few small fires had been allowed to burn in recent years.

The 1988 fires changed that. It was clear land managers needed to revisit their fire policies. Maybe, just maybe, control at all costs wasn't the best way forward.

———

Douglas firs can live over a thousand years, eventually towering two or even three hundred feet above the forest floor. Their cones are soft, softer than the pinecones I grew up with on the East Coast, and have little tufts on their edges that look like a mouse's back legs and tail poking out as it scurries inside the cone. After fires, these cones seed damp areas of charred forest, ready to take root in the open sunlight.

Douglas firs are one of the most common trees in the forests near where I live, but they aren't the only ones. Western and mountain hemlocks grow among them, their treetops drooping slightly like the tip of a wizard's hat, their petite cones only about the size of your thumb. Several kinds of firs—true firs, unlike the Douglas fir, which is an imposter—grow here. And, of course, there are the cedars, those sweet, evergreen-smelling behemoths with gently peeling bark.

I'm biased, of course, but I don't think there's a better smelling place than a forest in the Northwest, the fir and spruce and cedar and pine and damp earth mixing together into a scent that wafts around you as you hike. And in every corner of the forest, you'll find something living—a mushroom, a squirrel, an elk, a stand of fresh berries, a lichen wisping off the trees like an old man's beard. Without the trees, you'd have none of this.

Amy Harwood, a forestry activist, once told me that in the Northwest, trees take as long to die as they do to live. A Douglas fir scorched by fire may burst from the heat, much of its giant height plummeting to the ground. But that trunk on the forest floor decays, becoming home to fungi and insects and microbes. The part of it left standing will weather away, concealing feasts for woodpeckers and creating hiding places for birds' nests and more. In the sunny expanse the treefall opens, shrubs and saplings will take root.

We get so used to thinking linearly that forest fire seems simply tragic—there was a healthy stand of trees and now it's gone. But we forget this planet of ours works in cycles. After a fire, life begins again. It just looks a little different.

———

Growing up I felt different, but I also more or less fit in, made new friends relatively easily even if I never felt particularly close to anyone. I didn't know the few queer kids at my high school very well—or I knew them, but I was a little bit afraid of them. I didn't understand how they could be so brave, so outspoken. I wasn't ready to let people in to see that part of me.

It took years in college for me to find other queer friends, and even then I still mostly made friends with straight people. And it was the same when I moved to Seattle—I knew the queer people were somewhere, but I just couldn't seem to find them. I'd spent so long sunk into a world of grief and sadness that I didn't know how to *be* in the world. I felt rooted to one spot, scorched still.

I had always delved into queer history because it was easier to spend my time in what was known, in what I could get from books. History was safer, a knowable quantity, and I knew how to mourn. My shyness and awkwardness felt impossible to face.

I like to think now that learning my history, tethering myself to the queer past and all the people we've lost, eventually made it easier for me to build a new community. I like to think that I didn't just waste those years swamped in sadness. Instead, I hope it taught me how to live through the grief, how to move forward while still holding to a forest larger than myself. It taught me how to transform.

———

Fire is a normal, even essential, part of life in the West. But we've forgotten that, and it's putting us, and our homes, at risk.

The last few years, we've been seeing more and more destructive fires all along the West Coast. In addition to the Gorge fire, summer 2017 saw deadly fires throughout California: flames swept through Northern California in October, burning more than 350 square miles of Napa, Sonoma, and other counties, destroying towns and neighborhoods—homes—people assumed would always remain standing. The next year, the Pacific Northwest was largely spared, but California saw some of its worst fires in history. The Carr Fire destroyed sixteen hundred structures and killed eight people, while the Camp Fire destroyed more than twelve thousand structures and killed at least eighty-five people, devastating entire towns. In 2020, two enormous fires, the Lionshead and Beachie Creek, burned not far from Portland, destroying sixteen hundred structures, killing five people, and blanketing the city in smoke so thick we could barely see the house across the street for a week and couldn't go outside safely without wearing a respirator.

From far away, it all sounds so abstract, so unreal. But people have lost everything that tethers themselves to home—their houses, their landscapes, their family. Entire lives obliterated, burnt away. When Leslie and I drove through the burn zone the winter after the Lionshead and Beachie Creek fires, we saw hull after hull of burned-out houses, their charred skeletons the only evidence that they had once been homes.

In trying to make sense of this destruction, researchers have coined the phrase "wildland–urban interface," a human-made ecotone of sorts, one laden with risk rather than opportunity. Essentially, we're building further and further into areas prone to wildfire. Cities are expensive, and they're crowded; move to the edge of an urban area and you can have privacy, more affordable housing, open land, recreation opportunities, and more. From 1990 to 2000, the total wildland–urban interface land in the United States increased by almost 20 percent, and it continues to grow. But with wildland comes wildfire.

No one should lose their home, their life, to wildfire. But we also have to remember that building a home somewhere doesn't take the nature part of that landscape away. Our home becomes part of the landscape, not separate from it.

———

The fingers of climate change seem to creep in everywhere once you know where to look, and wildfires are no exception. With climate change comes warming—days oddly temperate and dry in winter and scorching hot in summer. That warming melts snowbanks and strips the moisture out of soil and undergrowth, leaving it primed and ready to burn. That's why the summer of 2017 could be such a bad year in the Northwest, fire-wise: we had record snowfall, sure, but the warm, sunny summer weather started around May rather than the usual July. So by August and September, that record snowpack had melted off and the fields of wildflowers had dried to a brittle brown. It was only a matter of time before the fires ignited.

Climate change will have less obvious effects, too, like changing weather patterns that bring the winds that make wildfires so spreadable and also so hard to fight. The way we're sprawling into the fire lands doesn't help either.

I'm reminded of a meme I've seen going around the internet dozens of times: one person remarks on "shark-infested waters," and another responds, "THEY LIVE THERE." It's the same thing here. Fires don't encroach on us; we encroach on them.

————

Ultimately, the teen who set the Gorge on fire—whose name still hasn't been released—wasn't tried as an adult. But he did receive a whopping punishment. In May 2018, he was sentenced to pay more than $36 million in restitution to the Eagle Creek fire victims, to write 152 apology letters to people affected by the fire, and to serve 1,920 hours of community service with the US Forest Service and five years of probation.

If you blithely light seventy-five miles of public forest on fire, if you damage a treasured resource and threaten people's homes, you deserve to be punished. You deserve to have to reassure your community that you are sorry and you won't go right back out and set off more fireworks and start more fires. But $36 million is a sum that that kid will never, ever, in his wildest dreams, be able to pay off. It's an amount so large it's hard to even understand just how much money that is. Does $36 million teach him that the world around him is to be treasured and respected, or just instill more disdain within him?

I believe he should be punished. But I also believe in restorative justice, and I don't think this is it.

After Leslie and I spent hike after hike talking about this kid, raging at all he had taken, our anger started to shift into a sadness, a frustration. Why were we once again pointing our finger at one person when our whole system felt broken?

This kid had grown up in a world with a changing climate altered by his parents, his grandparents, his great-grandparents. His ancestors and their communities had created a forest that was on the cusp of going up in flames. And he'd been taught by our culture that the right way to be a teen boy is to be destructive and blow shit up—that pyromania is just boys being boys, and that people will always be around to clean up his mess.

If we are going to punish the person who lit the match, we also have to look to the people who provided the kindling. Our forests are more prone to wildfire because we've made them that way. If we're going to point a finger at this teen, we also need to be pointing one at ourselves.

———

Robin Wall Kimmerer writes that "until we can grieve for our planet we cannot love it—grieving is a sign of spiritual health. But it is not enough to weep for our lost landscapes; we have to put our hands in the earth to make ourselves whole again."

I carry so much grief with me. Grief for my queer community, those I never knew and those who had to keep on living once they lost the ones they loved. Grief for what we're doing to this planet and for our apparent inability to change the direction we're moving in. And grief for myself, for the sadness and loneliness I have felt and for all the time it feels like I have lost to it.

When the Gorge fire first raged, I felt as if I was splintering in two, overwhelmed by grief for the beautiful places I'd explored and called home. Just weeks earlier, I'd watched those first plumes of smoke rise near Mount Jefferson, and now, here was yet another fire. Here was the summer sun, rusty red, blocked out by smoke blown over the city; here was the air, rough like sandpaper. My whole world, it seemed, was on fire. It felt like it would never stop.

And yet it did: the autumn rains came, as they always do in the Northwest, and the fires went out. The highway reopened, though it may be years before some of the trails that burned are walkable again. Whole slopes may collapse, the ancient canyon shedding a layer of skin.

A year after the fire, a few days after I hiked through the Mother Lode burn zone, I walked part of the Pacific Crest Trail into the Eagle Creek burn. At first, it felt like I was on some desolate planet, the greenery lost to sepia tones. The trees had crisped from roots to crown, shiny black bark rising to rusty brown needles. Wind trickled its way between the trees, and trunks creaked like old skeletons. Everything seemed dead.

But as I looked closer, I could see it. Little flashes of green, shoots of wild rose and columbine, fledgling currants and lupine. All these plants taking root in the spaces that had been cleared for them. It was all so fragile and new.

I still grieve for the old forest, for my queer family who has been lost to silence. But the forest remains, creating space for possibilities, despite all our best efforts to convince ourselves that it is gone.

8

LEGACY

In 2013, an enormous patch of water off the coast of Oregon, Washington, Western Canada, and Alaska started heating up, as if some unseen hand had come in and started fiddling with the ocean thermostat. Data buoys around the region logged temperature anomalies, little blips of balmy water. Over several months, the water temperature kept increasing until the surface reached more than two degrees Celsius warmer than had ever been recorded.

While two degrees doesn't sound like a lot, most sea life depends on very specific temperature ranges to survive. Two degrees of warming beyond the typical maximum in a cold ocean basin can be a death sentence. Plus, at its peak, this area of warming was some 3.5 million square miles—larger than the contiguous United States.

For its amorphous shape and massive size, scientists termed it the Blob. The name was cute, even cheeky, but like the Blob of the 1958 movie, it was soon to bring disaster wherever it went.

First, animals began to show up where they weren't expected. A Mola mola, the heaviest bony fish in the world, showed up off the coast of

Alaska, far north of the tropical and temperate water the species typically prefers. Velella velella, a beautiful blue jellyfish-like creature that usually floats along the open ocean, piled up along West Coast beaches, pushed by abnormal winds. Pacific saury, a subtropical fish, were spotted in huge numbers off British Columbia.

And then there were the Cassin's auklets. The Cassin's auklet is a chunky black orb of a bird that is small enough to be cupped in your hands. Despite their stocky, awkward shape, auklets are graceful ballerinas underwater, able to dive 120 feet down in search of prey like krill, copepods, and small fish. They spend most of their lives at sea or nesting on small islands along the coast. But between October and December that year, thousands of dead Cassin's auklets washed ashore from California to British Columbia, tiny bird carcasses littering the wrack line. All told, nearly half a million of them died.

Normally, cold water rises from the seafloor to the surface when it hits the continental shelf off the West Coast, a process called upwelling. The deep, cold water is full of nutrients that have fallen through the water column, and the upwelling brings them to the surface. There, the nutrients fuel blooms of tiny algae called phytoplankton, which in turn feed copepods, krill, and other small animals. But the Blob stopped that upwelling in its tracks, and the warmer water fueled blooms of smaller, less-nutritious food. Stuck, essentially, with junk food, the Cassin's auklets starved.

And they weren't the only ones. Fin whales and sea otters died in abnormally high numbers in Alaska, their bodies floating ashore and rotting in the waves, where they were then scavenged by seabirds and other predators. And all along the California coast, malnourished sea lion pups hauled out on beaches and rocks, wavering on the edge of death, their wrinkled skin looking three sizes too large and their faces already ghostly.

The year before the Blob hit, I was living in Seattle, my first foray into the Pacific Northwest. I was in grad school for creative writing, and when I wasn't writing, I usually was off in the mountains hiking, getting to know my new home.

Instead of driving east into the Cascades one summer weekend, I took the ferry west to the Olympic Peninsula and camped not far from the ocean. Early in the morning, the tide was out, and I went down in that hazy dawn summer light to a huge expanse of rocky flats. Briny sea smell filled the air while waves crashed in the distance. Water glistened in the dips between craggy black rocks. I had almost stayed in my sleeping bag that morning, hiding from the early coastal chill, but now I was glad I hadn't. This was the first time I'd seen tide pools, and it was like stepping through a veil into another world.

On areas of rocky coastline, creatures live by the rising and falling of the tide. At high tide, the rocks are covered by water, and it's like any other shallow stretch of sea. But twice a day, the water recedes, leaving behind exposed stone and small puddles. Waves pound away at the fringes while gulls, oystercatchers, crows, and other birds descend to the rocks to find a meal. Animals and plants here have to be versatile, ready to spend hours at a time out of water and able to protect themselves.

Standing on the low cliffs above the tide pools, everything had seemed quiet and still. But as I walked along the rocks, I realized everything around me was stunningly alive. On the higher rocks, further from the ocean, mussels covered almost every inch, clacking their shells shut like a chorus line of castanets to keep moisture in. White dragon claws of gooseneck barnacles edged out small areas for themselves wherever they could.

I walked further toward the water, careful not to slip on the algae covering the rocks. Tiny hermit crabs skittered under seaweed the moment I spotted them. Purple sea urchins nestled themselves in nooks and sea anemones tucked in their tentacles while they waited for the waves to return and bring them a meal.

Ochre stars in dusty purple, butter yellow, pinkish red, and luminous orange clung to the bottoms of the stones that rose above the sea. Sunflower stars the size of dinner plates aggregated in large groups, each with too many arms to count. I had never known sea stars could be so common. I wanted to sit there and stargaze until the tide came back in.

Three years later, when I began working as a science writer, I came back to the Olympic Peninsula with a group of educators and naturalists. Again, I picked my way across fields of algae and mussels, careful not to crush anything beneath my heavy hiking boots. Again, I marveled at the barnacles encrusting the rocks, at tiny fish swimming in the pools. At first glance, everything looked familiar. But as I crept from pool to pool, I also began to realize that everything was shining with a peculiar purple hue. I was seeing a vast army of purple sea urchins.

There was also hardly a single sea star to be found. At first, we could find only two ochre stars, one of them deformed and blob-like at its edges. We kept looking and finally found three more, blood stars this time, one only the size of a fingernail and the other two not much bigger. And that was it.

We had seen the end result of what scientists eventually termed sea star wasting syndrome. In June 2013, a year after my first trip to the peninsula, researchers on the Olympic coast had begun to notice something odd: the ochre stars seemed to be melting. First, a small lesion would appear on one of the star's arms, then another and another. Within a day or so, the arm would look like it was softening, then simply fall off. A sea star can lose an arm and regrow it—a useful strategy for dealing with predators—but this was different. The sea star would keep deteriorating, losing arm after arm until in a matter of days it would be dead.

By that August, divers noticed sunflower stars oozing and shedding limbs in the deeper waters just north of Vancouver, British Columbia. By October and November, sea stars near Monterey and Seattle were falling apart. By the end of the year, sea stars were wasting away in Southern California and Alaska, and by summer 2014, Oregon and Mexico were seeing their sea stars disintegrate too.

Ochre stars and sunflower stars are considered keystone species in tidal and subtidal environments like tide pools and kelp forests. Algae-eating invertebrates like purple sea urchins can strip a kelp forest or tide pool bare if left unchecked. But with sea stars around, the ecosystem stays balanced: a sunflower star, for example, can engulf an entire sea urchin, digest it, and then spit out the hard shell and move on to the next urchin. In the absence of sea stars, the sea urchin population can explode.

Within months, purple spiky spheres will cover the area, chowing down on algae. In addition to upsetting the balance in tide pools, this can be especially devastating in kelp forests—urchins tend to consume kelp holdfasts, sending the kelp spinning off into the current. Lose the kelp, lose the habitat for all sorts of other species. Lose the kelp, and all you have left is an urchin barren.

So what caused this carnage? Evidence points to the Blob. Sea stars don't have lungs or gills; instead, they take in oxygen passively through their skin. If there's not enough oxygen in the water, they can't just breathe harder or faster like we can—they suffocate. And marine heat waves like the Blob bring a double whammy of oxygen depletion: warm water holds less oxygen than cold water, and it fuels blooms of algae that decompose, feeding bacteria that use up whatever oxygen is still around. Right now, the prevailing theory is that the sea stars suffocated. Without enough oxygen, they wasted away.

Ochre and sunflower stars, as it turns out, tend to be the first species to succumb to the disease. That's why I hardly saw any when I returned to the pools. And without them, the way is cleared for sea urchins, a multitude of purple spikes sprawling all along the coast.

———

You can't blame sea urchins for what happens in the absence of sea stars: they come to tide pools and kelp forests because that's where their food is. They're just trying to survive, and if a sea star isn't around to eat them, all the better for the urchins.

Before they make it to the kelp forest or the tide pools, sea urchins spend months as larvae floating along ocean currents, wandering wherever the waves bring them, hopefully finding some food along the way and avoiding becoming someone else's meal. In its larval form, a sea urchin looks like a translucent cluster of pine needles, with spiny arms that can grab minuscule particles of phytoplankton. As the months pass, the urchin larva adds more arms. It spends months in this form, floating and growing, floating and growing, until it settles to the seafloor to become the angry ball we know so well.

Sea urchins are just one of the many ocean creatures that spend their early days or months as larvae drifting on the currents—sea stars, mussels, even octopuses move through larval stages. It's an essential time before a drastic metamorphosis that will change their shape forever.

This metamorphosis is what gives larvae their name. In Latin, one meaning of the word larva is *mask*. In 1768, Carl Linnaeus, father of all taxonomy, used the word to refer to juvenile forms of animals that look markedly different from their adult selves. In essence, larvae wear a false face, obscuring their final form.

But the word has another meaning: ghost, or demon. And while that's not the meaning Linnaeus was referencing, it's equally apt. In addition to being tiny, many oceanic larvae are translucent, ectoplasmic, so you can't see them unless you're really, really looking for them. Jump in the ocean, and you may not even notice you're surrounded by ghosts.

———

A number of animals we now consider charismatic were once demonized—orcas, for example. Many Indigenous cultures on the West

Coast have long respected them, but in Europe and later, in settler America, orcas were seen as pests at best. Pliny the Elder described them as "an enormous mass of flesh armed with teeth," and Western opinion didn't change much for centuries: orcas were considered bloodthirsty killers that competed with fishermen for catches.

Institutions like SeaWorld, which put on shows with orcas, changed our minds. Within the span of a few decades, orcas went from vilified to beloved. But the very thing that changed our minds about them may have spelled doom for the orcas that live throughout Puget Sound and the Salish Sea. These orcas—members of J, K, and L pods—are generally known as the southern residents, in contrast to the northern residents that live off the coast of British Columbia. Beyond being informally the mascots of the Pacific Northwest, the southern residents have a dubious claim to fame: they are the pods that Shamu and many other famous performing orcas came from.

In 1965, an orca from one of the southern residents' neighbors was found caught in a salmon net and sold to the now defunct Seattle Marine Aquarium where he was displayed as "Namu." More than one hundred thousand people came to see him in his first month at the aquarium. Though he died within a year, he set off a craze, and it didn't take long for aquarium operators to come to Puget Sound in search of more orcas.

In the early 1960s, there were roughly 120 southern resident orcas across J, K, and L pods. But by the mid-1970s, more than fifty of these orcas—almost half the population—had been captured or killed. Families of orcas were chased around Puget Sound and corralled in nets so that the younger ones could be separated from their parents. In one particularly horrible event on August 8, 1970, a family of whales from L pod were chased into Whidbey Island's Penn Cove by men in boats armed with explosives. The men surrounded the panicked whales with nets so they couldn't escape, then worked to separate the juveniles from their mothers. Seven were ultimately captured, but before that could happen, at least four others drowned, unable to stay afloat once they were tangled in the nets. "Stuff happens," said one diver later about their deaths. That diver and others later filled the young whales' carcasses with rocks to sink them and hide the evidence.

In the late 1970s, orca capture was banned in the Pacific Northwest, but the southern residents' numbers have never recovered. There's no doubt they're carrying with them the memory of what happened: Orcas are social, communicative animals that have strong matriarchal lines. They learn from one another. These orcas go through every day surviving the trauma that comes with watching their family be kidnapped and murdered in front of them, and losing the matriarchs who were the keepers of their culture.

––––––

Among the things we've already lost due to climate change and our generally neglectful view of the planet is a whole lot of history. Some of it is necessary: as we live our lives, we take up space, and some impacts are inevitable, however much we might prefer to tread lightly. But when we knock down buildings to create more efficient homes, we lose neighborhoods' histories; when we create new roads or expand the living space of cities, we pave over archaeological sites. As we've warmed the planet, rising sea levels and melting permafrost threaten to destroy thousands of historical burial grounds and other cultural sites of the Indigenous Peoples of the Circumpolar North; coastal sites all around the world are similarly at risk.

Western imperialism eradicated entire peoples and cultures, and their stories and languages are lost to the world, or survive in a curtailed, often hidden manner. In our quest for glory—for gold, for land, and now for oil—we've lost not only the culture-keepers but also physical records that might help us understand what we've lost.

After generations of cultural genocide—conquistador-led massacres, the mission system in California, residential schools throughout North America—some Indigenous cultures are slowly rebuilding, drawing from the knowledge that remains among the elders who survived. But it's much harder to build than it is to destroy. And if a group is entirely eradicated, like some Indigenous cultures in the now United States or, likely soon, southern resident orcas, that culture may be gone forever.

———

Southern resident culture depends on salmon—specifically Chinook salmon. Chinook are the largest species of Pacific salmon and historically have been found everywhere from Southern California all the way up to Alaska north of the Bering Strait. But today, the salmon are hard to find.

Salmon spend much of their adult lives at sea, but when it's time to breed, they make their way up rivers to the place they were born. They swim upstream against the current, making the leaps over rapids and waterfalls they're so famous for. But the longer they spend in fresh water, the more their saltwater bodies deteriorate. When they reach their traditional spawning grounds, they have just enough strength left to release their egg or sperm, then die a few days later. New fry make their way back to the ocean where the cycle begins again.

In the Northwest, we have some of the cheapest, least carbon-intensive energy in the nation, largely because of hydropower. Huge dams dot the Columbia and Snake rivers, two of the rivers most important to Chinook salmon. These hydroelectric dams generate an enormous amount of energy without burning fossil fuels—the Bonneville Dam alone, just a short way upriver from Portland, produces enough power for roughly nine hundred thousand homes.

But the dams also halt the traditional meandering flow of rivers, and these huge structures make it harder for fish like salmon to make their way to their spawning grounds. Even with fish ladders, only a small percentage of fish make it upstream, and these numbers pale in comparison to the number of salmon that traveled through the Columbia before it was known as the Columbia, when it was N'chi-Wana. These fish have dwindled in number because we make it nearly impossible for them to get to their spawning grounds, but for other reasons too. Pesticides and sediment from mining operations have poisoned and choked the water, while fewer salmon even get to the river in the first place because we've overharvested them. And climate change is warming the streams salmon spawn in, reducing the availability of oxygen in the water and making it harder for salmon fry to survive.

All this means that there's far less prey for the southern resident orcas. An orca can eat more than 300 pounds of salmon every day, so they feel every dip of the population. We disrupted their families once before, and now, we're doing it yet again.

———

The Bonneville Dam, the first blockade to Chinook salmon making their way up the Columbia River, is about halfway to many of the trails Leslie and I visit in the Columbia River Gorge. Usually, we drive past it without giving it much thought, but on the way home from a hike a few years ago, we pulled off the highway to watch the sunset. It was November, and moody streaks of gray clouds had replaced the endless blue sky of the summer. The setting sun had illuminated the sky in blazing oranges and reds, and it seemed a shame to just drive down the highway when all that was going on.

We'd never been to the dam, but signs said there was a little park there, so we pulled in and made our way past the security fences and industrial structures. We got out of the car and stood quietly on a cliff overlooking the Columbia and let the colors swim around us.

We'd been there a few minutes when a splash and a snuffling noise made us look down. We scanned the river. Nothing was there. Then, there, another snuffling, and out of the corner of my eye, I saw a small, brown, dog-like face pop out of the water. I squinted in the evening light. A sea lion? Impossible.

But it turns out, yes, it was a sea lion, miles from its home, upstream in strange fresh water. California and Steller sea lions journey more than one hundred miles up the river to hunt the fish that gather at the base of the fish ladders. The sea lions are highly effective predators, picking off thousands of salmon and steelhead below the dam each year.

Though numerous, these sea lion snacks are generally less than 5 percent of the fish that make their way up the fish ladders annually. But still, it sets up an impossible conflict for resource managers: sea lions are protected under the Marine Mammal Protection Act, but then again, salmon are protected under the Endangered Species Act and are a crucial fishery species to boot. The salmon fishery supports millions of dollars in income, and thousands of jobs. And just as they have formed the basis of Chinook culture for millennia, today they are part of the identity of the entire Northwest—"Salmon Nation," as Portland nonprofit Ecotrust refers to it. The mountain woods I so love depend on salmon: more than fifty species of mammal, bird, and fish feed on salmon and their eggs, and when the adult salmon die after spawning, they serve as fertilizer for the trees around them.

So while the sea lions are a protected species, so are the salmon, and the sea lion population is much healthier. That math means federal and state agencies are far more inclined to focus on the fish. The Oregon Department of Fish and Wildlife and the Port of Astoria at the mouth of the Columbia tried non-lethal hazing for a few years, using rubber bullets, small explosives, and even a fake orca in attempts to scare them off. But sea lions are smart and tenacious. They learned pretty quickly to avoid the people and the explosives, and they weren't particularly fooled by the orca, which sank anyway. Finally, the state got permission from the federal government to "remove" the sea lions—by which they generally meant kill.

We look at the sea lions eating the salmon and think we see the whole problem: get rid of the sea lions and the salmon will proliferate. If we widened our gaze, though, we might see the structure we placed so heavily in their path.

———

Up the Columbia River from the Bonneville Dam is the Dalles Dam, another enormous edifice spanning the river. Hydroelectric dams get built where water moves swiftly, so they can harness the water's kinetic energy as it falls and use it to generate electricity. The Dalles Dam is no exception: it's built over Celilo Falls, a series of rapids and waterfalls.

For more than eleven thousand years, Celilo Falls, or Wyam, was home to the oldest known continually inhabited settlement in North America, known as Nix lui dix. People would gather there to fish, trade, feast, and share information. Before the dam was built, great wooden platforms stood over the falls where fishermen gathered with dipnets to catch salmon making their way up the river. According to the Yakama, Umatilla, Warm Springs, and Nez Perce peoples, the salmon were the first gift given by the Creator to humans, and water, the home to the salmon, was the second.

Unlike its mostly languid cousin, the Mississippi, the Columbia is an unruly river. It twists and coils from its source high in the Canadian Rockies all the way to the coast, flowing north in British Columbia, turning back south into Washington, then finally setting a course generally to the west and forming the border between Washington and Oregon until it reaches the Pacific Ocean. Early settlers and white explorers—Lewis and Clark—found steep, impassable falls and roiling rapids that impeded their movements and overturned their canoes or forced lengthy portages. Between shifting sandbars and unpredictable currents and floods, the river could be deadly to those who would attempt to navigate it. Once white people began to settle the river's banks, they learned that those frequent floods were also liable to destroy homes and farmland constructed along its banks.

But that unruliness also created an opportunity. Over its course, the Columbia drops an average of a little over two feet per mile—a slope almost four times steeper than the Mississippi—and in many places it drops precipitously over falls like Celilo. Capture the energy carried by that falling water and you can power cities.

In the early- to mid-twentieth century, we did just that. Faced with the need to power growing metropolises like Seattle and Portland and to control the channel of the river to irrigate local agriculture, we began to dam the river. In the 1950s, our eyes turned to Celilo Falls. On March 10, 1957, the floodgates of the new dam closed to fill the new reservoir. Hours later, the falls were gone, and Wyam was lost to the waves.

In 2018, I took part in a residency that brought a group of writers and artists to what was once Wyam and Nix lui dix to see the petroglyphs that lined the cliffs, stories and messages left by Indigenous ancestors. Ed Edmo guided us to some of the few that remain in place. The most striking one is a face etched high upon a cliff, the eyes large and staring: Tsagaglalal, She Who Watches. According to the Wishram, Edmo told us, she was a chief who was changed into rock by Coyote so she could watch over her people forever.

Tsagaglalal watches forever, but behind a locked fence to protect her from spray paint and bullets. Many of the other petroglyphs are gone, covered by the waters that rose over Wyam when the dam was put in. The stories they contain have vanished, perhaps permanently. A few were rescued, and they rest, out of context, at nearby Columbia Hills State Park.

They, too, lie behind a fence, watched by a security camera, because time and time again they have been vandalized.

———

Here's one more fact: the three pods of southern resident orcas are among the most contaminated wild animals on Earth.

The Chinook salmon that the southern residents prey on live most of their lives in coastal waters, particularly in Puget Sound. There, they accumulate all sorts of toxins from urban runoff and industrial pollutants. Eat as much as these orcas do, and the chemicals begin to add up, building up over time in the orcas' blubber.

As a result, southern residents carry a huge load of what are known as persistent organic pollutants—compounds that have stuck around in our environment even after being banned and phased out decades ago. DDT, for one, and PCBs, polychlorinated biphenyl, a widely used coolant that turns out to be a carcinogen. Oh, and PBDE, polybrominated diphenyl ether, a flame retardant that has been shown to reduce fertility in humans. Even though we've restricted the use of all those compounds, we're still letting tons of new, questionably safe compounds wash into the ocean every day: of the eighty thousand or so synthetic chemicals in use today, only about sixteen hundred—2 percent—have been thoroughly assessed for toxicity. We assume the chemicals we use are safe until proven otherwise instead of testing them *before* we use them and release them into the world.

So now, not only are the southern resident orcas starving, they're also toxic.

In times of starvation, orcas, like most other animals, burn fat to survive. But that fat is exactly where they've been storing pollutants, and those chemicals then get released into their system in a double whammy of malnourishment and poison. Mother orcas also use their blubber to produce milk, so the youngest, most vulnerable orcas are getting a concentrated dose of what we've done.

————

It's not just orcas. A year after the Blob began, adult California sea lions began stranding along the California coastline, confused and disoriented. Some were dangerously thin, too addled to hunt. Others suffered seizures, their snouts held directly up toward the sky as if they were in a trance, their heads moving in a jerky, slow figure eight.

The culprit was domoic acid, a toxin that can cause neurological issues. During warm weather, domoic-acid-producing algae bloom on the ocean surface. From there, the toxin makes its way up the food chain, increasing in concentration each time: small zooplankton consume the algae, then fish consume the zooplankton, then larger fish consume those fish. By the time sea lions come in, there is enough domoic acid concentrated in their prey fish that it can cause permanent brain damage. Some sea lions don't survive it.

This time around, the Blob caused the ocean warming that led to the harmful algal bloom that killed the sea lions. But next time, it could be something else. For decades, the ocean has been operating as a heat sink, absorbing the majority of the warming we've caused by burning fossil fuels. As the ocean warms, algae expand their range, migrating toward formerly cold water nearer to the poles. Fueled by sunlight, these algae populations explode, causing blooms. And some of these algae produce toxins like domoic acid. So, thanks to climate change—thanks to us—this kind of thing may just be the new normal.

————

Domoic acid is naturally produced by algae. But most of the toxins in our environment are petrochemicals, derived from petroleum—oil and natural gas—and most have been made in the last century. Specifically, the boom in the creation of synthetic chemicals happened around World War II, as weapons and tools were developed for the war and afterward repurposed for lawn care, house cleaning, agricultural pesticides, and more. Over two human generations, the production of synthetic organic chemicals increased a hundred times.

My grandparents lived long, healthy lives: my mom's mom died a month shy of her ninetieth birthday, and my other grandparents lived well into their nineties. They grew up in a world reeling from the Great War, polluted by burning coal, and full of people struggling to survive the Depression, but they also grew up without the particular load of chemicals I almost certainly carry in my body.

The Baby Boomer generation—my parents' cohort—and my grandparents' generation before them are getting a significant amount of flak these days, deservedly so, for creating and exacerbating the conditions of climate change. While humans have been impacting the planet for thousands of years, the amount of carbon in our atmosphere and the average global temperature both began to skyrocket around 1950. These generations let chemicals go untested, dammed the Columbia and other rivers, and ignored scientists' persistent warnings that carbon dioxide was heating the planet. In their refusal to listen, they left the younger generations with one hell of a mess to clean up.

But has my generation done any better? Millennials like me are in our thirties now. Most of us are just waking up to the realities of climate change and environmental destruction, and few of us are in a position to do anything about it or willing to make the kind of changes that we need. What kind of world are we leaving behind?

The Haudenosaunee Confederacy holds the Seventh Generation Principle as one of its central tenets. It's all about taking "into consideration those who are not yet born but who will inherit the world." The Earth we live on is borrowed from future generations, and every decision we make must be with those descendants in mind. Other Indigenous Peoples throughout the world hold similar ideas as central to their cultures and their communities.

We are tied to the past and the future through our kin networks. Valerie Segrest, a Muckleshoot educator and advocate for food sovereignty, describes bringing her ancestors with her wherever she goes: "I always think," she said on the podcast *All My Relations,* "am I making my ancestors proud in these decisions I'm making or not?"

But unlike Segrest, I am not Indigenous. Some of my ancestors were refugees, yes, but many of them were settlers, coming to this continent to make money and claim what they saw as theirs—hardly caring for the land as if it were family. I have to ask a different question, or at least, I seek a different answer. If we descendants of settlers are doing things right, our ancestors should be spinning in their graves.

———

In 2018, a fleeting miracle happened: after several miscarriages, one orca in J pod known as Tahlequah gave birth to a living female calf. She was the first live southern resident calf in three years.

But the calf was so emaciated that she couldn't stay afloat, and after half an hour, she died. For seventeen days after her baby's death, Tahlequah carried the calf, holding her to the surface at the tip of her nose. When Tahlequah had to rest, her family carried the calf aloft. When Tahlequah needed to eat, her family kept her fed. The media termed it a "tour of grief," and that's exactly what it looked like, a whale and her family trying to show us all the horrors we have wrought.

Around the same time, one of Tahlequah's family members, Scarlet, started looking so thin that her ribs showed through her blubber. She regularly lagged up to a mile behind her family while they searched for prey. In an attempt to save her, researchers from NOAA Fisheries, which has jurisdiction over marine mammals, delivered antibiotics and a dewormer via a dart. They hoped her starvation was the result of or exacerbated by some sort of infection, something that could be fixed with medication. They also released farmed salmon in front of her, hoping she would catch and eat something.

Ultimately, though, it was too late. Scarlet was last seen on September 7, and the following week her family was witnessed traveling without her. There was nothing left to do but assume she had died.

Delivering antibiotics one orca at a time won't save the southern residents. But realistically, the actions that would keep the southern residents alive aren't simple, and to be effective, we'd need to have started them a long time ago. We would need to tear out the dams and figure out another way to keep the lights on all over the Northwest, to keep our farms irrigated and our towns from flooding. We would need to test all the synthetic compounds in use across the Columbia watershed and stop using the ones that caused harm, and find a way to clean the remaining toxins out of Puget Sound. It would be a huge undertaking, and the southern residents probably don't have that much time to spare.

But that doesn't mean we shouldn't try. Rather, we *must*. Unless we do more all over the world, the southern residents will be just one group of animals among many we've destroyed.

———

I've written about a lot of difficult, ugly things that we've done as humans throughout this book. But writing and researching this chapter is the first time I've cried while I've written. I've held back tears so I could see to finish pages that I've started, only to walk away from a draft to get in bed and sob. Grief over what we've done to everything living on this planet dug in under my skin and sat there like slow poison.

Maybe I did it to myself. I didn't have to watch videos of orcas being captured for our entertainment, to hear their panicked chirps and squeals and see them struggle to free themselves from nets. I didn't have to look at images of sea lions, their skins loose and wrinkled from starvation.

But writer Anna Badkhen points out that we have a tendency to place our empathy with victims, to imagine we can feel their pain along with them. That's a good thing, sure, in that it makes us want to help the orcas and the sea lions and the climate refugees. But it also means we avoid seeing ourselves in the perpetrators' position. We refuse to see that our greed warmed the waters, that our greed made the catastrophic storms. We refuse to see ourselves as culpable.

When I watched the orca videos, looked at the sea lion photos, I felt horrible for the animals whose lives and homes we've ruined. But I was also faced with the fact that *we* are the ones who ruined them. To remind myself that I, that all of us, are at fault.

If we are to make our way forward in this world, to survive as a species and to keep the ecosystems we depend on and love intact, we are going to have to change the way we think of ourselves.

I want to learn to think of myself as an ecological being. I want to see my body as mountain, as river, as bay. Not just in some abstract, poetic sense, but—

When I drink water from the tap, I want to remember to taste the mountainside it flowed down. When I breathe in, I want to remember that the oxygen filling my lungs came from the ocean. When I eat, I want to remember that the food I taste came from the ground. I want to remember that food and water are more important than plastic, than air conditioning, than oil.

I want to learn to think of myself not as a noun, discrete and individual, but rather as a verb—a cog in broad, endless networks, always acting. I want us all to learn that. I want to be able to spread my arms out and know that I am part of that queer utopia, the thing we could strive for if not for our obsession with wealth.

We have to learn to grieve for this planet and everything we've done, to let ourselves cry for the orcas and the sea lions and the coyotes and the rivers and the mountains. And even through our tears, we have to learn to say that yes, *we* have done these unspeakable things, and yes, we must speak these things, and then we have to change. We have to see the beauty in this world and be willing to give up everything—our settler ways, our iPhones, our plane rides, our belief that it isn't worth trying—to keep it alive.

———

When I turned 30, the year before COVID-19 hit, Leslie surprised me with a trip to the Oregon coast. We gathered our queer community together just up the hill from the beach for a long weekend.

The tides were low in the afternoon, and we bundled up against the February cold, donned waterproof boots and raincoats, and crossed the beach to the rocky pools that stretched along the coastline. I breathed in briny ocean air and felt myself choke back both grief and anticipation, a chill welling up within my chest. I steeled myself to see nothing but urchins.

The urchins were there, splashes of purple staining the rocks—but so were chitons, and anemones, and algae, and tiny, colorful sea slugs.

And then, out of the corner of my eye, I saw it: a flash of orange wedged in against a rocky crag. A sea star.

Once I saw the first, the others seemed so obvious. There were maybe ten or twenty, a constellation of ochre stars regaining their place within the intertidal. The sunflower stars were nowhere to be seen—they are, perhaps, gone forever—but the ochre stars are making their way back in this fragile world.

The tide pools are an ecotone, that liminal space that lets us see the stitching between worlds. If I'd been able to take a boat out to sea, I might have seen orcas; if I'd gone up the coast to the mouth of the river, a few Chinook salmon would have gathered for their spring journey. Against our best efforts to take up all the space on this planet, some animals have edged out small areas of existence.

We stayed there at the tide pools until sunset, watching as the ocean glowed gold and the water puddled in the rocks shined like a mirror. The wind blew across the beach, and I imagined I could hear distant coyotes calling to each other like otherworldly spirits.

I want to know that in our future—next year, next century—those sounds won't turn to silence.

ACKNOWLEDGMENTS

Much of this book was written while I lived on the unceded and traditional lands of the Chinook and other peoples, though I first imagined its possibility while living on Piscataway and Nacotchtank land. Generations upon generations of Indigenous communities have lived in relation and care with the land in the places I lived in and visited while writing this book, and I am grateful to them. They should never have had their lands and waters taken from them. They deserve to have them back.

Unsettling was many years in the thinking and writing and revising, and so many people helped me along the way. I will do my best to name them all here, and I apologize to anyone I have missed.

Arielle Datz took a chance on *Unsettling* and tirelessly advocated for it to have a place in the world. Lisa Kloskin helped me see the book in a new light and hone it into its final shape. Huge thanks to the entire team at Broadleaf: Erin Gibbons, Elle Rogers, Emily Benz, Marta Smith, Jana Nelson, and Michele Lenger.

The Louisiana Universities Marine Consortium brought me to Terrebonne Parish for OCEANDOTCOMM; I am especially grateful to Virginia Schutte and Craig McClain. Michelle Barboza-Ramirez, Becca Burton, Melissa Cronin, M.B. Humphrey, Gabi Serrato Marks, Bethann Garramon Merkle, and all of the other OCEANDOTCOMM attendees were brilliant collaborators. Delaina LeBlanc was incredibly patient with this beginner birder and mist netter. The Pointe-au-Chien Tribe, especially Patty Ferguson-Bohnee and Theresa Dardar, welcomed me into their home, fed me, and shared their stories.

The Signal Fire residency and my fellow residents—especially Christina Catanese, Monica Vaughan, and Anne Greenwood—helped me reframe my relationship to forest fire. Ed Edmo generously shared stories of the Columbia River and invited us to visit Tsagaglalal.

The entire community of the LGBTQ Outdoor Summit taught me that queerness and the outdoors could share a place in my heart. I am grateful to the home that I have found there and for Hannah Malvin's friendship and tireless work making it happen.

Jami Hammer welcomed Leslie and me to the Indiana Coyote Rescue Center, even though it was Christmas week and about ten degrees and windy, and told us everything we could possibly want to know about coyotes. Thank you to Ares, Neegan, Artemis, and Orion: may you live long and cozy lives with plentiful canned peas.

Many scientists responded to my often-unprompted emails and patiently answered my questions. Thank you to Nyssa Silbiger, for sharing information about tide pools and ocean acidification; Craig Smith, for talking to me about whale falls; Nick Pyenson, for an unforgettable tour of the Smithsonian's fossil collection; Morgan Haldeman, for explaining plate tectonics and volcanism; Barbara King, Marc Bekoff, and Bill Lynn, for insights about animal grief and coyote emotions; Mark Pagani, for letting me sit in on his paleoclimates class; and Ian Hewson, for helping me untangle my understanding of sea star wasting syndrome.

Thank you to my many teachers over the years, who taught me to write, to read closely, and to be persistent: Lidia Yuknavitch, Maya Sonenberg, David Bosworth, Andrea Barrett, Jim Shepard, Karen Russell, Katie Kent, Sara Dubow, and Suzanne Doggett.

Thank you to friends and colleagues who read innumerable drafts, shared ideas, encouraged me, and graciously listened to my despair: my Body of the Book cohort, especially Rhea Wolf, Michelle Fredette, and Daniel Elder; Kristine Greive, brilliant friend, insightful editor, and fellow channeled scabland fan; Emily Maclary; Amber Marsh; Candise

Branum; Meghan Lake; Leah Lansdowne; Lorenzo Triburgo; Erin Bookout; and friends and colleagues at the NOAA Office of National Marine Sanctuaries, especially Dayna McLaughlin, Vernon Smith, and Grace Bottitta-Williamson, and at the Interagency Arctic Research Policy Committee—Sara Bowden, Nikoosh Carlo, Meredith LaValley, Hazel Shapiro, Sorina Stalla, Natasha Gamache, Larry Hinzman, Sandy Starkweather, and Danielle Stickman.

Endless, endless thanks to my family, who have believed in and supported me forever: my parents, Kathie and David, and Michael, Steven, Jessica, Casey, Amina, and Felix. Thank you also to Deb, Jeffrey, Jeff, and Megan, who have welcomed me as family with open arms.

Goose passed away when I had not quite finished the full draft of this book, and I miss her every day. Pigeon has stepped up like a champion, bringing me her toys when I've been working for too long, lying on the floor with me when I get tired, and being the best hiking buddy I could ask for.

And to Leslie: I truly believe I could not have done this without you—you believed in this book, and me, even when I didn't. Thank you. I love you.

SUGGESTED FURTHER READING

I am indebted to many scholars, activists, artists, and others who have been writing and thinking about issues of land, identity, climate change, and more for many years. While I cited their work where possible within the text of *Unsettling*, many books and other sources influenced my thinking without directly showing up on the page. What follows is an acknowledgement of sorts and a resource for anyone who would like to dive further into these topics.

Abbey, Edward. *Desert Solitaire.* New York: Ballantine, 1968.

Bagemihl, Bruce. *Biological Exuberance: Animal Homosexuality and Natural Diversity.* New York: St. Martin's Press, 1999.

Bechdel, Allison. *Fun Home.* Boston and New York: Mariner Books, 2006.

Beckoff, Marc. *The Emotional Lives of Animals.* Novato, California: New World Library, 2007.

Biss, Eula. *Notes from No Man's Land.* Minneapolis: Graywolf Press, 2009.

Bjornerud, Marcia. *Timefulness: How Thinking Like a Geologist Can Help Save the World.* Princeton, NJ: Princeton University Press, 2018.

Brennan, Summer. *The Oyster War: The True Story of a Small Farm, Big Politics, and the Future of Wilderness in America.* Berkeley: Counterpoint, 2015.

Bruso, Jenny. @unlikelyhikers. *Instagram.* https://www.instagram.com/unlikelyhikers/.

Chauncey, George. *Why Marriage? The History Shaping Today's Debate Over Gay Equality.* New York: Basic Books, 2004.

Clare, Eli. *Brilliant Imperfection: Grappling with Cure.* Durham, NC and London: Duke University Press, 2017.

————*Exile and Pride: Disability, Queerness, and Liberation.* Durham, NC: Duke University Press, 2015.

Cole, Luke W., and Sheila R. Foster. *From the Ground Up: Environmental Racism and the Rise of the Environmental Justice Movement.* New York and London: New York University Press, 2001.

Crimp, Douglas, and Adam Rolston. *AIDS Demographics.* Seattle: Bay Press, 1990.

Dietrich, William. *Northwest Passage: The Great Columbia River.* Seattle: University of Washington Press, 1995.

Duggan, Lisa. *The Twilight of Equality: Neoliberalism, Cultural politics, and the Attack on Democracy.* Boston: Beacon Press, 2004.

Dunbar-Ortiz, Roxanne. *An Indigenous People's History of the United States.* Boston: Beacon Press, 2014.

Edelman, Lee. *No Future: Queer Theory and the Death Drive.* Durham, NC: Duke University Press, 2004.

Egan, Timothy. *The Big Burn: Teddy Roosevelt & the Fire that Saved America.* Boston: Mariner Books, 2009.

Estes, Nick. *Our History Is the Future.* London/New York: Verso, 2019.

Finney, Carolyn. *Black Faces, White Spaces: Reimagining the Relationship of African Americans to the Great Outdoors.* Chapel Hill, NC: University of North Carolina Press, 2014.

Flores, Dan. *Coyote America: A Natural and Supernatural History.* New York: Basic Books, 2016.

France, David. *How to Survive a Plague: The Inside Story of How Citizens and Science Tamed AIDS.* New York: Alfred A. Knopf, 2016.

Franke, Katherine. *Wedlocked: The Perils of Marriage Equality.* New York: New York University Press, 2015.

Freedman, Eric. "In the Shadow of Death." *Earth Island Journal* Volume 32, Number 1 (Spring 2017): 42–46.

Fuller, Randall. *The Book that Changed America: How Darwin's Theory of Evolution Ignited a Nation.* New York: Viking, 2017.

Ghosh, Amitav. *The Great Derangement: Climate Change and the Unthinkable.* Chicago: University of Chicago Press, 2016.

Gumbs, Alexis Pauline. *Undrowned: Black Feminist Lessons from Marine Mammals.* Chico, CA: AK Press, 2020.

Houston, Pam. *Deep Creek: Finding Hope in the High Country.* New York: W.W. Norton & Company, 2019.

Hurley, Nat. "The Little Transgender Mermaid: A Shape-Shifting Tale." In *Seriality and Texts for Young People,* edited by M. Reimer et al. UK: Palgrave Macmillan, 2014. 258–280.

Irvine, Amy. *Desert Cabal: A New Season in the Wilderness.* Salt Lake City, Utah: Torrey House Press, 2018.

Jensen, Tori. "Women in the Fracklands: On Water, Land, Bodies, and Standing Rock." *Catapult,* January 3, 2017. https://catapult.co/stories/women-in-the-fracklands-on-water-land-bodies-and-standing-rock.

Kendi, Ibram X. *Stamped from the Beginning: The Definitive History of Racist Ideas in America.* New York: Nation Books, 2016.

Kimmerer, Robin Wall. *Braiding Sweetgrass: Indigenous Wisdom, Scientific Knowledge, and the Teachings of Plants.* Minneapolis: Milkweed Editions, 2013.

———*Gathering Moss: A Natural and Cultural History of Mosses.* Corvallis, OR: Oregon State University Press, 2003.

King, Barbara J. *How Animals Grieve.* Chicago: University of Chicago Press, 2013.

Klein, Naomi. *This Changes Everything: Capitalism vs. the Climate.* New York: Simon & Schuster, 2014.

Kolbert, Elizabeth. *The Sixth Extinction: An Unnatural History.* New York: Henry Holt, 2014.

———*Under a White Sky: The Nature of the Future.* New York: Crown, 2021.

LaDuke, Winona. *All Our Relations: Native Struggles for Land and Life.* Chicago: Haymarket Books, 1999.

————*To Be a Water Protector: The Rise of the Wiindigoo Slayers*. Halifax & Winnipeg: Fernwood Publishing, and Ponsford MN: Spotted Horse Press, 2020.

Laing, Olivia. *The Lonely City: Adventures in the Art of Being Alone*. New York: Picador, 2016.

————*To the River*. Edinburgh: Canongate Books, 2011.

Lopez, Barry. *Horizon*. New York: Alfred A Knopf, 2019.

————*Of Wolves and Men*. New York: Scribner, 1978.

Lopez, Barry, and Debra Gwartney, editors. *Home Ground: A Guide to the American Landscape*. San Antonio: Trinity University Press, 2006.

Lorde, Audre. *Zami: A New Spelling of My Name*. Berkeley: Crossing Press, 1982.

Love, Heather. *Feeling Backward: Loss and the Politics of Queer History*. Cambridge: Harvard University Press, 2007.

Macfarlane, Robert. *Underland: A Deep Time Journey*. New York: W. W. Norton, 2019.

Magnason, Andri Snær. *On Time and Water*. Translated by Lytton Smith. Rochester, NY: Open Letter, 2021.

McPhee, John. *The Control of Nature*. New York: Farrar, Straus & Giroux, 1989.

Miranda, Deborah A. *Bad Indians*. Berkeley, CA: Heyday, 2013.

Mortimer-Sandilands, Catriona. "Unnatural Passions?: Notes Toward a Queer Ecology." *Invisible Culture* Issue 9 (2005). https://ivc.lib .rochester.edu/unnatural-passions-notes-toward-a-queer-ecology/.

Mortimer-Sandilands, Catriona, and Bruce Erickson, editors. *Queer Ecologies: Sex, Nature, Politics, Desire*. Bloomington and Indianapolis: Indiana University Press, 2010.

Muir, John. *Wilderness Essays*. Layton, Utah: Gibbs Smith, 2015.

Muñoz, José Esteban. *Cruising Utopia: The Then and There of Queer Futurity*. New York: New York University Press, 2009.

Nagakyrie, Syren. @disabledhikers. *Instagram*. https://www.instagram .com/disabledhikers/.

Nelson, Maggie. *The Argonauts*. Minneapolis: Graywolf Press, 2016.

Nixon, Rob. *Slow Violence and the Environmentalism of the Poor*. Boston: Harvard University Press, 2013.

Orange, Tommy. *There There*. New York: Alfred A. Knopf, 2018.

Owen, David. *Where the Water Goes: Life and Death Along the Colorado River*. New York: Riverhead Books, 2017.

Pasternak, Judy. *Yellow Dirt: A Poisoned Land and the Betrayal of the Navajos*. New York: Free Press, 2011.

Pyenson, Nick. *Spying on Whales: The Past, Present, and Future of Earth's Most Awesome Creatures*. New York: Viking, 2018.

Ray, Sarah Jacquette. *The Ecological Other: Environmental Exclusion in American Culture*. Tucson: University of Arizona Press, 2013.

Reeves, Randall R., Brent S. Stewart, Phillip J. Clapham, and James A. Powell. *National Audubon Society Guide to Marine Mammals of the World*. New York: Alfred A. Knopf, 2002.

Roman, Joe et al. "Whales as marine ecosystem engineers." *Frontiers in Ecology and the Environment* Volume 12, Issue 7 (2014): 377–385. doi: 10.1890/130220.

Roy, Arundhati. *Walking with the Comrades*. New York: Penguin, 20–12.

Rybczynski, Witold. *A Clearing in the Distance: Frederick Law Olmsted and America in the 19th Century*. New York: Touchstone, 1999.

Savoy, Lauret Edith. *Trace: Memory, History, Race, and the American Landscape*. Berkeley, CA: Counterpoint Press, 2015.

Shaffer, Leah. "Systemic Racism Affects Wildlife, Too: A Q&A With An Urban Ecologist." *Discover*, Feb 11, 2021. https://www.discovermagazine.com/planet-earth/systemic-racism-affects-wildlife-too-a-q-and-a-with-an-urban-ecologist.

Smith, Craig R. "Bigger Is Better: The Role of Whales as Detritus in Marine Ecosystems." In *Whales, Whaling, and Ocean Ecosystems*, edited by James A. Estes at al., 284–299. Berkeley, CA: University of California Press, 2006.

Smith, Craig R. et al. "Whale-Fall Ecosystems: Recent Insights into Ecology, Paleoecology, and Evolution." *Annual Review of Marine Science* 7 (2015): 571–596. doi: 10.1146/annurev-marine-010213-135144.

Solnit, Rebecca. *The Encyclopedia of Trouble and Spaciousness*. San Antonio: Trinity University Press, 2014.

———*The Faraway Nearby*. New York: Penguin, 2013.

———*A Field Guide to Getting Lost*. New York: Penguin, 2005.

———*Hope in the Dark*. Chicago: Haymarket Books, 2016.

———*Savage Dreams: A Journey into the Hidden Wars of the American West*. Oakland, CA: University of California Press, 2000.

———*Wanderlust: A History of Walking*. New York: Penguin, 2001.

Steingraber, Sandra. *Living Downstream: An Ecologist's Personal Investigation of Cancer and the Environment*. 2nd edition. Cambridge, MA: Da Capo Press, 2010.

Stevens, J. Richard. *Captain America, Masculinity, and Violence: The Evolution of a National Icon*. Syracuse, New York: Syracuse University Press, 2015.

Sinopoulos-Lloyd, Pınar, and So Sinopoulos-Lloyd. @queernature. *Instagram*. https://www.instagram.com/queernature/.

Talaga, Tanya. *All Our Relations: Finding the Path Forward*. Canada: House of Anansi Press, 2018.

Taylor, Dorceta. *The Rise of the American Conservation Movement: Power, Privilege, and Environmental Protection*. Durham and London: Duke University Press, 2016.

Taylor, Sunaura. *Beasts of Burden: Animal and Disability Liberation*. New York: The New Press, 2017.

Treuer, David. *The Heartbeat of Wounded Knee: Native America from 1890 to the Present*. New York: Riverhead Books, 2019.

Treuer, David. "Return the National Parks to the Tribes." *The Atlantic*, April 12, 2021. https://www.theatlantic.com/magazine/archive/2021/05/return-the-national-parks-to-the-tribes/618395/.

Tuck, Eve, and K. Wayne Yang. "Decolonization is not a metaphor." *Decolonization: Indigeneity, Education & Society* Volume 1, Number 1 (2012): 1–40.

Varela, Jolie. @indigenouswomenhike. *Instagram.* https://www.instagram.com/indigenouswomenhike/.

Warner, Michael. *The Trouble with Normal: Sex, Politics, and the Ethics of Queer Life.* Cambridge, Massachusetts: Harvard University Press, 1999.

Washuta, Elissa. *White Magic.* Portland, OR: Tin House, 2021.

Washuta, Elissa, and Theresa Warburton, editors. *Shapes of Native Nonfiction.* Seattle: University of Washington Press, 2019.

Wilbur, Matika, and Adrienne Keene. *All My Relations.* Podcast. https://www.allmyrelationspodcast.com/.

Williams, Joy. *Ill Nature: Rants and Reflections on Humanity and Other Animals.* New York: Vintage Books, 2001.

Wojnarowicz, David. *Close to the Knives: A Memoir of Disintegration.* New York: Vintage Books, 1991.

Wulf, Andrea. *The Invention of Nature: Alexander von Humboldt's New World.* New York: Vintage Books, 2015.

NOTES

INTRODUCTION

viii *why live a life that's just one thing?* Maggie Nelson, *The Argonauts* (Minneapolis: Graywolf, 2016), 74.

viii *fewer whales began showing up in Hawai'i each year* Elizabeth Weinberg, "Humpback Whales are Navigating an Ocean of Change." *NOAA Office of National Marine Sanctuaries*, September 2018. https://sanctuaries.noaa.gov/news/sep18/humpback-whales-navigating-an-ocean-of-change.html.

x *the United Nations released a report* IPCC, 2021: Summary for Policymakers. In: *Climate Change 2021: The Physical Science Basis. Contribution of Working Group I to the Sixth Assessment Report of the Intergovernmental Panel on Climate Change* [Masson-Delmotte, V., P. Zhai, A. Pirani, S. L. Connors, C. Péan, S. Berger, N. Caud, Y. Chen, L. Goldfarb, M. I. Gomis, M. Huang, K. Leitzell, E. Lonnoy, J.B.R. Matthews, T. K. Maycock, T. Waterfield, O. Yelekçi, R. Yu and B. Zhou (eds.)]. Cambridge University Press. In Press.

I. GRAVEYARD

3 *more than twenty billion miles of arteries, veins, and capillaries* Nick Pyenson, *Spying on Whales: The Past, Present, and Future of Earth's Most Awesome Creatures* (New York: Viking, 2018), 117.

4 *more small organisms than in any other recorded habitat below a thousand meters* Craig R. Smith, "Bigger Is Better: The Role of Whales as Detritus in Marine Ecosystems," in *Whales, Whaling, and Ocean Ecosystems*, ed. James A. Estes et al. (University of California Press, 2006), 286.

4 *a humpback whale skeleton may contain the weight of two small cars in fats* Smith, "Bigger Is Better," 284.

4 *The bones may last these organisms for up to a century* Pyenson, *Spying on Whales*, 78.

5 *More than four hundred species* Smith, "Bigger Is Better," 289.

7 *responsible for some 71 percent of greenhouse gas emissions* Climate Accountability Institute, "The Carbon Majors Database: CDP Carbon Majors Report 2017," CDP, July 2017.

7 *In 2014, the average American released more than twice as much carbon dioxide* Lisa Friedman, Nadja Popovich, and Henry Fountain, "Who's Most Responsible for Global Warming?" *New York Times*, April 26, 2018, https://nyti.ms/2Ke6sjg.

8 *Nearly three million whales were killed* Daniel Cressey, "World's Whaling Slaughter Tallied: Commercial Hunting Wiped out Almost Three Million Animals Last Century," *Nature* (March 11, 2015), http://www.nature.com/news/world-s-whaling-slaughter-tallied-1.17080.

8 *a single percentage of their historical population in the Southern Hemisphere* Joe Roman et al., "Whales as Marine Ecosystem Engineers," *Frontiers in Ecology and the Environment* (2014): 1, DOI: 10.1890/130220.

8 *their numbers dropping from 327 thousand in 1904* Line Bang Christensen, *Marine Mammal Populations: Reconstructing Historical Abundances at the Global Scale* (Vancouver, Canada: University of British Columbia Fisheries Centre, 2006).

8 *just a couple thousand in the early part of this century* "Population (Abundance) Estimates," International Whaling Commission, accessed October 3, 2021, https://iwc.int/estimate.

8 *literally millions fewer tons of carbon are captured* Andrew J. Pershing et al., "The Impact of Whaling on the Ocean Carbon Cycle: Why Bigger was Better," *PLoS ONE* (August 26, 2010), doi: 10.1371/journal.pone.0012444.

9 *Whale fall expert Craig Smith* Craig Smith, personal communication, June 28, 2018; Amanda Mascarelli, "Dead Whales Make for an Underwater Feast," *Audubon Magazine,* November–December 2009, http://www.audubon.org/magazine/november-december-2009/dead-whales-make-underwater-feast.

9 *a balena may be four acres large with eyes as big as fifteen men* Joseph Nigg, *Sea Monsters: A Voyage Around the World's Most Beguiling Map* (Chicago and London: The University of Chicago Press, 2013), 50.

9 *"unless they can save themselves by ropes thrown forth of the ship, are drown'd"* Nigg, *Sea Monsters*, 112.

10 *In Indigenous whaling communities* Bathsheba Demuth, "Arctic Energy Before Petroleum," The Arctic Institute, January 12, 2021, https://youtu.be/H4UsSVl2ubI.

12 *"I long for you, yes . . . loved so much by me as you"* Jackie Wullschlager, "Introduction," in *Fairy Tales*, Hans Christian Andersen, trans Tiina Nunnally (New York: Viking, 2004), xxvii.

12 *"gloriously . . . although he never knew"* Hans Christian Andersen, "The Little Mermaid," in *Fairy Tales*, trans Tiina Nunnally (New York: Viking, 2004), 84–85.

13 *Howard Ashman built the character after a sketch* Nicole Pasulka and Brian Ferree, "Unearthing the Sea Witch," Hazlitt, January 14, 2016, https://hazlitt.net/longreads/unearthing-sea-witch.

15 *scientists have found stone harpoon points and other whaling technologies* John C. "Craig" George and John R. Bockstoce, "Two Historical Weapon Fragments as an Aid to Estimating the Longevity and Movements of Bowhead Whales," *Polar Biology* (December 21, 2007), doi: 10.1007/s00300-008-0407-2. http://www.north-slope.org/assets/images/uploads/George_bockstoce_08-Polar-Biol_Yankee_whale_bomb_fragment_FINAL.pdf.

2. ICE

18 *two-thousand-year-old forests* Laura Nielsen, "A Forest Revealed Under Glacial Ice," Frontier Scientists, September 25, 2013, https://frontierscientists.com/2013/09/forest-revealed-under-glacial-ice/.

18 *six-thousand-year-old collections of Viking artifacts* Jason Daley, "Norway's Melting Glaciers Release Over 2,000 Artifacts," *Smithsonian Magazine*, January 26, 2018, https://www.smithsonianmag.com/smart-news/2000-artifacts-pulled-edge-norways-melting-glaciers-180967949/.

18 *ten-thousand-year-old hunting weapons* Jim Scott, "CU Researcher Finds 10,000-Year-Old Weapon," *Colorado Arts and Sciences Magazine*, June 21, 2010, https://www.colorado.edu/asmagazine/2010/06/21/cu-researcher-finds-10000-year-old-weapon.

18 *thirty-thousand-year-old seeds* Svetlana Yashina et al., "Regeneration of Whole Fertile Plants from 30,000-y-Old Fruit Tissue Buried in Siberian Permafrost," *PNAS* (March 6, 2012), doi: 10.1073/pnas.1118386109.

18 *a mummified body dating back to the Copper Age* "The Iceman," South Tyrol Museum of Archaeology, https://www.iceman.it/en/the-iceman/.

24 *The water escaped the lake at nearly sixty miles an hour* William Dietrich, *Northwest Passage: The Great Columbia River* (Seattle: University of Washington Press, 1995), 128–30.

26 *"spacious, prairie-like glacier"* John Muir, "The Discovery of Glacier Bay," in *Wilderness Essays* (Layton, Utah: Gibbs Smith, 2015), 25.

26 *retreated a massive thirty-one miles between 1892 and 2005* "Repeat Photography of Alaskan Glaciers," USGS, accessed via the Internet Archive Wayback Machine, https://web.archive.org/web/20160304190750/https://www2.usgs .gov/climate_landuse/glaciers/repeat_photography.asp.

26 *rise by up to eight feet by the end of the century* Jeffrey Payne et al., "Chapter 8: Coastal Effects," in *Impacts, Risks, and Adaptation in the United States: Fourth National Climate Assessment, Volume II*, eds. D.R. Reidmiller et al. (Washington, DC: US Global Change Research Program), https://nca2018.globalchange .gov/chapter/8/.

26 *the Solomon and Marshall Islands, Palau, Samoa, Tuvalu, and others* Leo Berthe, Denis Chang Seng, and Lameko Asora, "Multiple Stresses, Veiled Threat: Saltwater Intrusion in Samoa," in Samoa Conference III: Opportunities and Challenges for a Sustainable Cultural and Natural Environment (Apia, Samoa: National University of Samoa, August 25–29, 2014), http://samoanstudies .ws/wp-content/uploads/2015/03/Leo-Berthe-Dennis-Chang-Seng-and -Lameko-Asora.pdf.

26 *it was warm enough in Antarctica* J.E. Francis et al., "100 Million Years of Antarctic Climate Evolution: Evidence from Fossil Plants," in *Antarctica: A Keystone in a Changing World*, eds. A.K. Cooper et al., Proceedings of the 10th International Symposium on Antarctic Earth Sciences (Washington, DC: The National Academies Press, 2008), doi: 10.3133/of2007-1047.kp03.

26 *alligators lived in the Arctic* Jaelyn J. Eberle and David R. Greenwood, "Life at the Top of the Greenhouse Eocene World—A Review of the Eocene Flora and Vertebrate Fauna from Canada's High Arctic," *GSA Bulletin* (January 1, 2012), doi: 10.1130/B30571.1.

27 *will likely feel the effects of sea level rise less* Carol Rasmussen, "Glacial Rebound: The Not So Solid Earth," NASA, Aug 26, 2015, https://www.nasa.gov/feature /goddard/glacial-rebound-the-not-so-solid-earth; "Why Is Sea Level Rising Faster in Some Places Along the US East Coast Than Others?" Woods Hole

Oceanographic Institution, December 19, 2018, https://www.whoi.edu/press
-room/news-release/why-is-sea-level-rising-higher-in-some-places-along-u
-s--east-coast-than-others.

28 *"Queerness is not yet here . . . a horizon imbued with potentiality"* José Esteban
Muñoz, *Cruising Utopia: The Then and There of Queer Futurity* (New York: New
York University Press, 2009), 1.

30 *the number of people who are "alarmed" about climate change* Anthony Leiserowitz
et al., "Global Warming's Six Americas in 2020." *Yale Program on Climate
Change Communication* (October 10, 2020), https://climatecommunication
.yale.edu/publications/global-warmings-six-americas-in-2020/.

30 *The first paper about the harmful effects of carbon dioxide* Svante Arrhenius, *On the
Influence of Carbonic Acid in the Air upon the Temperature of the Ground* (1896)
republished in *The Global Warming Reader,* ed. Bill McKibben (2011).

34 *"strong . . . friends in olden times' vibes"* Kaila Hale-Stern, "I Am Begging
You Not to Write Articles that Refute a Fictional Character's Sexuality,"
The Mary Sue, April 30, 2021, https://www.themarysue.com/bucky-barnes-
sexuality-kari-skogland-variety-interview/.

34 *fanfics available on the Archive of Our Own* ToastyStats. "[Fandom stats] Biggest
Fandoms, Ships, and Characters on AO3 (2021)," Archive of Our Own, February 6, 2021, https://archiveofourown.org/works/29249172.

34 *ice cores can take us back* "Ice cores and climate change," British Antarctic
Survey, March 1, 2014, https://www.bas.ac.uk/data/our-data/publication/ice
-cores-and-climate-change/.

35 *Ice core records have been obtained from tropical and subtropical mountains* Lonnie
G. Thompson et al., "Tropical Glaciers, Recorders and Indicators of Climate
Change, are Disappearing globally," *Annals of Glaciology* (September 14, 2017),
doi: 10.3189/172756411799096231.

35 *some sixteen hundred years' worth of ice* Marcia Bjornerud, *Timefulness: How
Thinking Like a Geologist Can Help Save the World* (Princeton NJ: Princeton
University Press, 2018), 130–31; L.G. Thompson et al., "Annually Resolved Ice
Core Records of Tropical Climate Variability over the Past ~1800 Years." *Science*
340 (May 24, 2013), doi: 10.1126/science.1234210.

35 *at least 86 percent of the ice cover* Thompson et al. "Tropical glaciers, recorders and
indicators of climate change, are disappearing globally."

35 *Elsewhere, we've lost more recent records* Qianggong Zhang, Shichang Kang, Paolo Gabrielli, Mark Loewen, and Margit Schwikowski. "Vanishing High Mountain Glacial Archives: Challenges and Perspectives." *Environmental Science & Technology* Volume 49, Issue 16 (2015): 9387–10254. doi: 10.1021/acs.est.5b03066.

36 *By 2030, they may all be gone* "World of Change: Ice Loss in Glacier National Park," NASA Earth Observatory, https://earthobservatory.nasa.gov/world-of-change/Glacier.

3. WILDS

42 *"an area where . . . a visitor who does not remain"* United States Congress. *Public Law 88-577, Section 2.* Wilderness Act. September 3, 1964.

45 *"the hard contests...all they hold dear"* Teddy Roosevelt, "The Strenuous Life" (speech, The Hamilton Club, Chicago, April 10, 1899).

45 *not considered to be a place for women* Dorceta Taylor, *The Rise of the American Conservation Movement: Power, Privilege, and Environmental Protection* (Durham and London: Duke University Press, 2016), 85.

46 *their food stores burned* David Treuer, "Return the National Parks to the Tribes," *The Atlantic* (April 12, 2021), https://www.theatlantic.com/magazine/archive/2021/05/return-the-national-parks-to-the-tribes/618395/.

46 *as if it had truly been a two-sided battle* Rebecca Solnit, *Savage Dreams: A Journey into the Hidden Wars of the American West* (Oakland, CA: University of California Press, 2000).

46 *90 percent of the Ahwahneechee inhabitants* "Surviving Communities," Yosemite National Park, National Park Service, https://www.nps.gov/yose/learn/historyculture/surviving-communities.htm.

46 *"No foot seems to have neared it"* Quoted in Solnit, *Savage Dreams*, 220.

47 *it wasn't just the land that was taken* Isaac Kantor, "Ethnic Cleansing and America's Creation of National Parks," *Public Land and Resources Law Review* 28 (2007): 47.

48 *"A bay is a noun only . . . a flock of baby mergansers"* Robin Wall Kimmerer, *Braiding Sweetgrass: Indigenous Wisdom, Scientific Knowledge, and the Teachings of Plants* (Canada: Milkweed Editions, 2013), 55; for a similar example in Lakotayapi, see Nick Estes, *Our History is the Future* (London/New York: Verso, 2019), 9.

49 *By the time of European contact* Roxanne Dunbar-Ortiz, *An Indigenous People's History of the United States* (Boston: Beacon Press, 2014), 22.

49 *abided by the Doctrine of Christian Discovery* Estes, *Our History is the Future*; David Treuer, *The Heartbeat of Wounded Knee: Native America from 1890 to the Present* (New York: Riverhead Books, 2019); Dunbar-Ortiz, *An Indigenous Peoples' History of the United States.*

50 *"The tribes of Indians . . . to leave the country a wilderness."* United States Supreme Court. *Johnson & Graham's Lessee v. McIntosh.* Justia, February 27, 1823, https://supreme.justia.com/cases/federal/us/21/543/.

50 *"our manifest destiny . . . unjust to ourselves"* John O'Sullivan, "Annexation." *United States Magazine and Democratic Review* 17, no. 1 (July–August 1845): 5–10.

51 *"Invasion . . . broke the plains"* Rebecca Solnit, *Savage Dreams*, 312.

56 *"betrays . . . an inhabited place uninhabited"* Eula Biss, "No Man's Land," in *Notes from No Man's Land* (Minneapolis: Graywolf Press, 2009), 159.

57 *"We all might do well . . . chooses to inhabit the world"* Lauret Edith Savoy, *Trace: Memory, History, Race, and the American Landscape* (Berkeley, California: Counterpoint Press, 2015), 87.

58 *"Vanish . . . They continued to speak"* Savoy, *Trace*, 72–73.

60 *Dghelay Ka'a, "big mountain"* Alan Boraas, "We know it's not McKinley, but is Denali the Right Name for our Mountain?" *Anchorage Daily News*, June 28, 2015, https://www.adn.com/commentary/article/we-know-its-not-mckinley-denali-right-name-our-mountain/2015/06/28/.

62 *"Decolonization is not a metonym for social justice"* Eve Tuck and K. Wayne Yang, "Decolonization Is Not a Metaphor." *Decolonization: Indigeneity, Education & Society* 1, no. 1 (2012).

4. SURVIVAL

63 *a lone coyote trotted his way* Bob Sallinger, "God's Dog Rides MAX: Urban Coyotes," *Columbia Slough Watershed*, 293–6, https://www.fws.gov/oregonfwo/externalaffairs/outreach/Documents/WildRead/Coyotes%20small.pdf.

68 *more than fifteen metric tons of carbon dioxide* "CO2 Emissions (Metric Tons per Capita)," The World Bank, 2018, https://data.worldbank.org/indicator/en.atm.co2e.pc?view=map.

69 *up to two thousand coyotes are estimated to live in and around the city* "How Many Coyotes Are in a Pack?" Urban Coyote Research Project, https://urbancoyoteresearch.com/faq/how-many-coyotes-are-pack.

69 *more than eighteen hundred sightings were reported* "Research Snapshot," Portland Urban Coyote Project, https://www.portlandcoyote.com/research.html.

69 *Coyote is known for causing the Missoula floods* Ed Edmo (story told August 31, 2018).

70 *"From the 1880s . . . the very shortest period"* Dan Flores, *Coyote America: A Natural and Supernatural History* (New York: Basic Books, 2016), 84.

71 *Bounties set by state legislatures* Flores, *Coyote America*, 88.

71 *This went on for decades* Dan Flores, "Stop Killing Coyotes," *New York Times*, August 11, 2016, https://nyti.ms/2aYa8DV.

72 *the bison are Pte Oyate, the buffalo nation* Estes, *Our History is the Future*, 109.

72 *settlers turned to the bison* Estes, *Our History is the Future*, 78.

72 *"the violent expression . . . what we fear they may do"* Barry Lopez, *Of Wolves and Men* (New York: Scribner, 1978), 140.

74 *Lesbian, gay, and bisexual people are twice as likely* "Sexual Orientation and Estimates of Adult Substance Use and Mental Health: Results from the 2015 National Survey on Drug Use and Health," Substance Abuse and Mental Health Services Administration, October 2016, https://www.samhsa.gov/data/sites/default/files/NSDUH-SexualOrientation-2015/NSDUH-SexualOrientation-2015/NSDUH-SexualOrientation-2015.htm#topofpage.

74 *trans people are almost four times more likely* Jonathon W. Wanta et al., "Mental Health Diagnoses Among Transgender Patients in the Clinical Setting: An All-Payer Electronic Health Record Study," *Transgender Health* 4, no. 1 (2019), doi: 10.1089/trgh.2019.0029.

74 *Lesbian, gay, and bisexual youth* Laura Kann et al., "Sexual Identity, Sex of Sexual Contacts, and Health-Related Behaviors Among Students in Grades 9–12—United States and Selected Sites, 2015," *Surveillance Summaries* 65, no. 9: 1–202, doi: 10.15585/mmwr.ss6509a1.

74 *trans youth have it even worse* Tracy A. Becerra-Culqui et al., "Mental Health of Transgender and Gender Nonconforming Youth Compared with Their Peers," *Pediatrics* (2018), doi: 10.1542/peds.2017-3845.

75 *have all killed off queer characters* Marie Lyn Bernard (Riese), "All 210 Dead Lesbian and Bisexual Characters on TV, and How They Died," *Autostraddle*,

March 11, 2016, last modified May 1, 2021, https://www.autostraddle.com /all-65-dead-lesbian-and-bisexual-characters-on-tv-and-how-they-died -312315/?all=1.

75 *AIDS was the leading cause of death* A.R. Jonsen and J. Stryker, eds., *The Social Impact of AIDS in the United States* (Washington, DC: National Academies Press, 1993), Chapter 9, https://www.ncbi.nlm.nih.gov/books/ NBK234564/.

75 *contained nineteen hundred panels and weighed three and a half tons* David France, *How to Survive a Plague: The Inside Story of How Citizens and Science Tamed AIDS* (New York: Alfred A. Knopf, 2016), 295.

76 *have been seen as far south as Panama* Allison Hody et al., "Canid Collision— Expanding Populations of Coyotes (*Canis latrans*) and Crab-Eating Foxes (*Cerdocyon thous*) Meet up in Panama," *Journal of Mammalogy* 100, no. 6 (December 19, 2019), https://doi.org/10.1093/jmammal/gyz158.

78 *The subsequent ecological and cultural disaster* Treuer, *The Heartbeat of Wounded Knee*, 65–66.

79 *we've killed about five hundred thousand coyotes a year* Dan Flores, *Coyote America*, 16.

80 *Researchers have documented coyotes patrolling* Marc Beckoff, *Animals Matter* (Boston: Shambala Publications, 2007), 20–21.

81 *In 2019, a trans woman was assaulted in Portland* Blair Stevnick, "Queer Portlanders Shaken by Reports of Anti-LGBTQ Violence," Blogtown - Portland Mercury, February 19, 2019, https://www.portlandmercury.com/blogtown /2019/02/18/25935639/queer-portlanders-shaken-by-reports-of-anti-lgbtq-violence.

81 *Twenty states still have no laws on the books* "Non-Discrimination Laws," Movement Advancement Project, accessed July 8, 2021, http://www.lgbtmap.org/ equality-maps/non_discrimination_laws.

81 *Halfway into 2021* Wyatt Ronan, "2021 Officially Becomes Worst Year in Recent History for LGBTQ State Legislative Attacks as Unprecedented Number of States Enact Record-Shattering Number of Anti-LGBTQ Measures into Law," Human Rights Campaign, May 7, 2021, https:// www.hrc.org/press-releases/2021-officially-becomes-worst-year-in-recent -history-for-lgbtq-state-legislative-attacks-as-unprecedented-number-of -states-enact-record-shattering-number-of-anti-lgbtq-measures-into-law.

81 *In 2020, at least forty-four trans and gender-nonconforming people* "Fatal Violence Against the Transgender and Gender Non-Conforming Community in 2020," Human Rights Campaign, https://www.hrc.org/resources/violence-against-the-trans-and-gender-non-conforming-community-in-2020.

81 *kill an estimated 2.4 billion birds each year* Scott R. Loss, Tom Will, and Peter P. Marra, "The Impact of Free-Ranging Domestic Cats on Wildlife in the United States." *Nature Communications* 4, no. 1396 (January 29, 2013), doi: 10.1038/ncomms2380., https://3pktan2l5dp043gw5f49lvhc-wpengine.netdna-ssl.com/wp-content/uploads/2015/09/Loss_et_al._2013-Impacts_Outdoor_Cats.pdf.

5. EXTINCTION

88 *recent computer models suggest that the thylacine's jaw was too weak* M.R.G. Attard et al., "Skull Mechanics and Implications for Feeding Behaviour in a Large Marsupial Carnivore Guild: the Thylacine, Tasmanian Devil and Spotted-Tailed Quoll," *Journal of Zoology* (November 9, 2011), doi: 10.1111/j.1469-7998.2011.00844.x.

91 *she gives them all a waking dream* Tamora Pierce, *Emperor Mage* (New York: Random House, 1995), 194.

92 *"civilized . . . the lowest savages"* Ibram X. Kendi, *Stamped from the Beginning: The Definitive History of Racist Ideas in America* (New York: Nation Books, 2016), 210.

92 *"survival of the fittest . . . inferior races"* Kendi, *Stamped from the Beginning,* 210.

93 *the evolutionary "missing link"* Carolyn Finney, *Black Faces, White Spaces: Reimagining the Relationship of African Americans to the Great Outdoors* (Chapel Hill, NC: University of North Carolina Press, 2014), 41.

93 *About six thousand animal species* "Zoos and Aquarium Statistics," Association of Zoos & Aquariums, last modified April 2021, https://www.aza.org/zoo-and-aquarium-statistics.

95 *at most nineteen vaquitas remained* Armando M. Jaramillo-Legorreta et al., "Decline Towards Extinction of Mexico's Vaquita Porpoise (*Phocoena sinus*)," *Royal Society Open Science* (July 31, 2019), https://doi.org/10.1098/rsos.190598.

95 *a group of experts from nine different countries* VaquitaCPR. https://www.vaquitacpr.org/.

100 *One-third of all reef-building corals . . . One-sixth of all birds* Elizabeth Kolbert, *The Sixth Extinction: An Unnatural History* (New York: Henry Holt and Company, 2014), 17.

102 *238 ecoregions within a series of "biogeographic realms"* D.M. Olson and E. Dinerstein, "The Global 200: Priority Ecoregions for Global Conservation." *Annals of the Missouri Botanical Garden* 89, no. 2 (2002): 199–224.

103 *Erle Ellis and Navin Ramankutty asked the same questions* Erle C. Ellis and Navin Ramankutty, "Putting people in the map: anthropogenic biomes of the world," *Frontiers in Ecology and Environment* 6, no. 1 (2008): 439–47, https://doi.org/10.1890/070062.

103 *Humans have altered an estimated 97 percent of land* Andrew J. Plumptre et al., "Where Might We Find Ecologically Intact Communities?" *Frontiers in Forests and Global Change* (April 15, 2021), doi: 10.3389/ffgc.2021.626635.

105 *"life and death sometimes hangs on an acknowledgement of personhood"* Eli Clare, *Brilliant Imperfection: Grappling With Cure* (Durham, NC and London: Duke University Press, 2017), 30.

105 *More than a thousand errant sightings have been recorded* Barry W. Brook et al., "Extinction of the Thylacine," BioRxiv, January 19, 2021, https://doi.org/10.1101/2021.01.18.427214.

6. River

107 *A fire sparked not far from the interstate* John Hartig, "Great Lakes Moment: Five decades since the infamous Rouge River fire," Great Lakes Now, October 2019, https://www.greatlakesnow.org/2019/10/rouge-river-fire-anniversary-great-lakes-moment/.

109 *had made the Potomac "disgraceful"* Lyndon B. Johnson, "Remarks at a Meeting of the Water Emergency Conference," August 11, 1965 in *1965 (In Two Books): Containing the Public Messages, Speeches, and Statements of the President, Book 2*, Lyndon B. Johnson, Public Papers of the Presidents of the United States, 870, https://quod.lib.umich.edu/p/ppotpus/4730960 .1965.002/342?page=root; size=100;view=image;q1=Water+Emergency+Conference11.

111 *"In divers places . . . their heads above the water"* "Smith's First Voyage," Captain John Smith Four Hundred Project, Sultana Projects, http://www.johnsmith400 .org/journalfirstvoyage.html.

112 *"What I love most... same river twice" Pocahontas.* Directed by Mike Gabriel and Eric Goldberg, Walt Disney Pictures, 1995.

113 *mistrustful of Indigenous people* John Smith, *A Map of Virginia: With a Description of the Countrey, the Commodities, People, Government and Religion* (Wisconsin Historical Society, 2003), 150–1, http://www.americanjourneys.org/pdf/AJ -075.pdf.

113 *more than a little self-aggrandizing* John Smith. *A True Relation by Captain John Smith, 1608* (Wisconsin Historical Society, 2003), http://www .americanjourneys.org/pdf/AJ-074.pdf.

116 *turned some 217 square miles of land and marsh into open water* John A. Barras, "Land Area Changes in Coastal Louisiana After Hurricanes Katrina and Rita," in *Science and the Storms: the USGS Response to the Hurricanes of 2005*, ed. G.S. Farris et al. (US Geological Survey Circular 1306, 2007), 97–112, https://pubs .usgs.gov/circ/1306/pdf/c1306_ch5_b.pdf.

117 *the Bureau of Indian Affairs requires a tribe to trace its ancestry* Sara Sneath, "Louisiana Tribes say Federal Recognition will Help to Face Threat of Climate Change," NOLA.com, July 26, 2018, https://www.nola.com/expo/news /erry-2018/07/449c2f22d39490/louisiana-tribes-say-federal-r.html.

117 *"at the crossroads of adaptation or extinction"* Patty Ferguson-Bohnee, "The Impacts of Coastal Erosion on Tribal Cultural Heritage," *Forum Journal* (Summer 2015): 62–63, https://ssrn.com/abstract=2742326.

120 *"The Colorado River is the most endangered river . . . we do to the other"* Natalie Diaz, "The First Water is the Body," *Postcolonial Love Poem* (Minneapolis, MN: Graywolf Press, 2020).

122 *three times as much lawn space* Christina Milesi et al., "Mapping and Modeling the Biogeochemical Cycling of Turf Grasses in the United States," *Environmental Management* (2005), doi: 10.1007/s00267-004-0316-2.

122 *US spending on lawn care in 2015* Kristin Runge, "How Much do Americans Spend per Capita on Lawn and Garden Care?" Madison.com, April 28, 2017, https://madison.com/wsj/business/how-much-do-americans-spend-per-capita-on-lawn-and-garden-care/article_ef2654fa-0b36-511c-b16d-3e9bfd34f21b.html; "GSP (Current US$) – Iceland," The World Bank, https:// data.worldbank.org/indicator/NY.GDP.MKTP.CD?locations=IS.

123 *one hundred million trees died* USDA Office of Communications, "New Aerial Survey Identifies more than 100 Million Dead Trees in California,"

US Forest Service, November 18, 2016, https://www.fs.fed.us/news/releases/new-aerial-survey-identifies-more-100-million-dead-trees-california.

124 *Four hundred counties sought federal disaster relief* Sarah Almukhtar et al., "The Great Flood of 2019: A Complete Picture of a Slow-Motion Disaster," *New York Times*, September 11, 2019, https://www.nytimes.com/interactive/2019/09/11/us/midwest-flooding.html.

124 *more than a million acres of cropland* P.J. Huffstutter and Humeyra Pamuk, "More than 1 Million Acres of US Cropland Ravaged by Floods," Climate Signals, March 29, 2019, https://www.climatesignals.org/headlines/exclusive-more-1-million-acres-us-cropland-ravaged-floods.

124 *Tar sands mining requires 2.3 gallons of water* Naomi Klein, *This Changes Everything: Capitalism vs. the Climate* (New York: Simon & Schuster, 2014) 346.

124 *"People who live with water . . . it's profound."* Winona LaDuke, *To Be a Water Protector: The Rise of the Wiindigoo Slayers* (Halifax, Nova Scotia and Winnipeg, Manitoba: Fernwood Publishing; Ponsford, Minnesota: Spotted Horse Press, 2020), 2.

124 *"Our people . . . as long as we live."* LaDuke, *To Be a Water Protector*, 93.

124 *"Wherever they live . . . even die for it."* Klein, *This Changes Everything*, 347.

125 *Police from twenty-four counties and sixteen cities in ten different states joined local law enforcement to "manage"* LaDuke, *To be a Water Protector*, 107.

125 *coordinated military-style counterterrorism methods* Alleen Brown, Will Parrish, and Alice Speri, "Leaked Documents Reveal Counterterrorism Tactics Used at Standing Rock to 'Defeat Pipeline Insurgencies.'" The Intercept, May 27, 2017, https://theintercept.com/2017/05/27/leaked-documents-reveal-security-firms-counterterrorism-tactics-at-standing-rock-to-defeat-pipeline-insurgencies/.

127 *"a horizon imbued with potentiality"* Muñoz, 1.

7. FIRE

132 *There's an iconic photo* Timothy Rawles, "Picture from 1993 Reminds People of the Loss of Life due to AIDS." *San Diego Gay & Lesbian News*, November 28, 2017, https://sdlgbtn.com/causes/2017/11/28/picture-1993-reminds-people-loss-life-due-aids.

134 *"one of the finest remaining examples of temperate rainforest in the United States"* "Visiting the Hoh Rain Forest," Olympic National Park, National Park Service, https://www.nps.gov/olym/planyourvisit/visiting-the-hoh.htm.

134 *A single Sitka spruce in the Hoh Rainforest* Marin Hutten, Andrea Woodward, and Karen Hutten. "Inventory of the Mosses, Liverworts, Hornworts, and Lichens of Olympic National Park, Washington: Species List" in *US Geological Survey, Scientific Investigations Report 2005–5240* (2005), 55, https://pubs.usgs.gov /sir/2005/5240/sir20055240.pdf.

135 *thirty-two thousand cases had been reported in the United States* Douglas Crimp and Adam Rolston, *AIDS Demographics* (Seattle: Bay Press, 1990), 27.

135 *President Ronald Reagan didn't even say the word* German Lopez, "The Reagan Administration's Unbelievable Response to the HIV/AIDS Epidemic," *Vox* (Dec 1, 2016), https://www.vox.com/2015/12/1/9828348/ronald-reagan-hiv-aids.

135 *Their actions were also generally accompanied by graphic, often controversial posters and flyers* Crimp and Rolston, *AIDS Demographics*.

140 *the "idiot teen" who had set the fire* Katie Herzog, "Idiot Teen Faces Charges for Columbia Gorge Fire," *The Stranger* (October 20, 2017), https://www .thestranger.com/slog/2017/10/20/25484731/idiot-teen-faces-charges-for -columbia-gorge-fire.

141 *"Of all the foes . . . no other is so terrible as fire"* Timothy Egan, *The Big Burn: Teddy Roosevelt & the Fire That Saved America* (Boston: Mariner Books, 2009), 35.

141 *Fire was not new to the Bitterroot Mountains* Egan, *The Big Burn,* 111; Stephen W. Barrett and Stephen F. Arno, "Indian Fires in the Northern Rockies," in *Indians, Fire, and the Land in the Pacific Northwest* (Oregon: OSU Press, 1999).

142 *Pinchot changed his agency's priorities* Egan, *The Big Burn,* 241.

143 *Fifty-one separate fires raged within the park* "1988 Fires," Yellowstone National Park, National Park Service, https://www.nps.gov/yell/learn/nature/1988fires .htm.

144 *It was a rediscovery of sorts* David Edward Yarlott, "Historical Uses of Natural Resources: Transference of Knowledge in the Crow Indian Environment." (EdD diss., Montana State University, 1999), https://scholarworks .montana.edu/xmlui/bitstream/handle/1/8588/31762104226772. pdf;sequence=1.

147 *the total wildland-urban interface land in the United States increased by almost 20 percent* Susan M. Stein et al., *Wildfire, Wildlands, and People: Understanding and Preparing for Wildfire in the Wildland-Urban Interface—a Forests on the Edge report*, US Forest Service, Rocky Mountain Research Station, 10, https://www .fs.fed.us/openspace/fote/reports/GTR-299.pdf.

148 *"shark-infested waters . . . THEY LIVE THERE"* "Some Jackalope," Toothybabies, https://montereybayaquarium.tumblr.com/post/148856708288/some-jackalope -shark-infested-waters-me-cupping.

149 *"until we can grieve for our planet . . . make ourselves whole again"* Robin Wall Kimmerer, *Braiding Sweetgrass: Indigenous Wisdom, Scientific Knowledge, and the Teachings of Plants* (Canada: Milkweed Editions, 2013), 327.

8. Legacy

152 *Stuck, essentially, with junk food* Timothy Jones et al., "Massive Mortality of a Planktivorous Seabird in Response to a Marine Heatwave," *Geophysical Research Letters* (February 28, 2018), https://doi.org/10.1002/2017GL076164.

155 *Evidence points to the Blob* C.D. Harvell et al., "Disease Epidemic and a Marine Heat Wave are Associated with the Continental-Scale Collapse of a Pivotal Predator (*Pycnopodia helianthoides*)," *ScienceAdvances* 5, no. 1 (January 30, 2019), doi: 10.1126/sciadv.aau7042.

155 *the prevailing theory is that the sea stars suffocated* Ian Hewson, email message to author, August 2, 2021; Ian Hewson, "Explanation of Boundary Layer Oxygen Diffusion Limitation Hypothesis for Sea Star Wasting," Team Aquatic Virus at Cornell, August 24, 2020, https://team-aquatic-virus.com/2020/08/24 /explanation-of-boundary-layer-oxygen-diffusion-limitation-hypothesis -for-sea-star-wasting/; Citlalli Aquino et al., "Evidence for Boundary Layer Oxygen Diffusion Limitation as a Key Driver of Asteroid Wasting," bioRxiv, August 4, 2020, doi: 10.1101/2020.07.31.231365.

157 *"an enormous mass of flesh armed with teeth"* Pliny the Elder, *The Natural History*, trans. John Bostock and H.T. Riley (London: Taylor and Francis, 1855), http:// www.perseus.tufts.edu/hopper/text?doc=Perseus:text:1999.02.0137:book =9:chapter=5.

157 *More than one hundred thousand people came to see him* Evan Thompson, "From Pest to Pet to Icon: Our Evolving Relationship with Orcas," *HeraldNet*,

November 4, 2018, https://www.heraldnet.com/life/from-pest-to-pet-to-icon-our-evolving-relationship-with-orcas/.

157 *more than fifty of these orcas . . . had been captured or killed* "Southern Resident Orcas," Endangered Species Coalition, https://www.endangered.org/campaigns/southern-resident-orcas/.

158 *"Stuff happens"* Jeffrey Ventre, "Orca Stories – Penn Cove – Don Goldsberry of SeaWorld Inc." YouTube, uploaded July 29, 2012, 04:22, https://www.youtube.com/watch?v=iUlbZifjoqo.

160 *only a small percentage of fish make it upstream* "Chinook Salmon," Salish Sea, United States Environmental Protection Agency, https://www.epa.gov/salish-sea/chinook-salmon.

161 *"Salmon Nation"* "Salmon Nation," Ecotrust, https://ecotrust.org/publication/salmon-nation/.

162 *was home to the oldest known continually inhabited settlement in North America* Dietrich, *Northwest Passage*, 52.

162 *According to the Yakama, Umatilla, Warm Springs, and Nez Perce peoples* "We are all Salmon People," Columbia River Inter-Tribal Fish Commission, https://www.critfc.org/salmon-culture/we-are-all-salmon-people/.

164 *a flame retardant that has been shown to reduce fertility* K.G. Harley et al., "PBDE Concentrations in Women's Serum and Fecundability," *Environmental Health Perspectives* 118, no. 5 (May 2010): 699–704, doi: 10.1289/ehp.0901450.

164 *of the eighty thousand or so synthetic chemicals in use today* Sandra Steingraber, *Living Downstream: An Ecologist's Personal Investigation of Cancer and the Environment*, Second Edition. (Cambridge, MA: Da Capo Press, 2010), 102.

166 *the production of synthetic organic chemicals increased a hundred times* Steingraber, *Living Downstream*, 90–91.

166 *"into consideration those who are not yet born but who will inherit the world"* "Values," Haudenosaunee Confederacy, https://www.haudenosauneeconfederacy.com/values/.

167 *I always think . . . proud in these decisions I'm making or not?"* Matika Wilbur and Adrienne Keene, "Food Sovereignty: A Growing Movement," March 2, 2019, in *All My Relations*, podcast, Season 1, Episode 2, https://www.allmyrelationspodcast.com/post/ep-2-food-sovereignty-a-growing-movement.

169 *we have a tendency to place our empathy with victims* Anna Badkhen, "The Journey: The Writer in the World," conversation with Barry Lopez, Panel at the 2019 Portland Book Festival, https://literary-arts.org/event/the-journey -the-writer-in-the-world/.